CO_2 Laser Cutting

Springer
London
Berlin
Heidelberg
New York
Barcelona
Budapest
Hong Kong
Milan
Paris
Santa Clara
Singapore
Tokyo

John Powell

CO_2 Laser Cutting

Second Edition

With 103 Figures

Springer

Dr John Powell, DSc, DIC, CEng, CPhys

Technical Director, Laser Expertise Ltd., Unit H,
Acorn Park Industrial Estate, Harrimans Lane, Dunkirk,
Nottingham NG7 2TR, UK

Visiting Professor, Department of Materials Processing,
Lulea University of Technology, S-951 87 Lulea, Sweden

Academician, Engineering Academy of St. Petersburg, Russia

Cover illustration: Ch.5, Fig.7b. Nozzle with a movable lens.

ISBN 1-85233-047-3 2nd edition Springer-Verlag Berlin Heidelberg New York

ISBN 3-540-19786-9 1st edition Springer-Verlag Berlin Heidelberg New York
ISBN 0-387-19786-9 1st edition Springer-Verlag New York Berlin Heidelberg

British Library Cataloguing in Publication Data
Powell, John, 1955-
CO_2 laser cutting. - 2nd ed.
1. Laser beam cutting 2. Carbon dioxide lasers
I. Title
671.5'3
ISBN 1852330473

Library of Congress Cataloging-in-Publication Data
A catalog record for this book is available from the Library of Congress

Printed in Germany

First published 1993
Second edition 1998

Typeset by Best-Set Typesetter Ltd., Hong Kong
69/3830-543210 Printed on acid-free paper

Contents

Foreword

The laser has given manufacturing industry a new tool. When the laser beam is focused it can generate one of the world's most intense energy sources, more intense than flames and arcs, though similar to an electron beam. In fact the intensity is such that it can vaporise most known materials. The laser material processing industry has been growing swiftly as the quality, speed and new manufacturing possibilities become better understood. In the fore of these new technologies is the process of laser cutting. Laser cutting leads because it is a direct process substitution and the laser can usually do the job with greater flexibility, speed and quality than its competitors.

However, to achieve these high speeds with high quality considerable know how and experience is required. This information is usually carefully guarded by the businesses concerned and has to be gained by hard experience and technical understanding. Yet in this book John Powell explains in lucid and almost non-technical language many of these process wrinkles concerning alignment, cornering, pulsing, water jets, material properties, cutting speeds as well as tricks with surface coating and much much more. It is a book which managers and technicians in laser job shops and laser processing facilities would be foolish not to read.

John Powell has been working with lasers since the 1970s. He is one of the most experienced specialists in the subject of laser cutting today. He was awarded one of the early doctorates in the subject of laser material processing while working with me at Imperial College. He is now the Technical Director of one of the more successful laser job shops in the UK. Perhaps reading the book will tell the reader why his business is successful. He is also a visiting Professor to the University of Lulea in Sweden, an Academician of the Engineering Academy of St. Petersburg and a well known character at many laser conferences. Certainly the book helps the whole subject to take a step forward and not become constipated in secrecy. There is much more to learn about laser material processing for it is the application of a new energy source with the new horizons that must open for us. This book is a step on this long and exciting journey that has just begun. But apart from all that this book is a pleasure to read – so learn, enjoy and relax in John Powell's down to earth style!

W.M. Steen
James Bibby Professor of Engineering Manufacture,
Mechanical Engineering Department,
The University of Liverpool,
Liverpool L69 3BX

Introduction

This book is intended to be useful to anyone interested in the subject of CO_2 laser cutting. The subject itself is rather like car maintenance in that it involves a rather complex interrelationship between many individually simple components. At the risk of stretching the metaphor it is nevertheless true that there is no need to know anything about car maintenance in order to drive a car and, in the same way, there is no need to know anything about the mechanism of laser cutting in order to run a CO_2 laser cutting machine. In both cases, however, any information you do have will undoubtedly be useful at some time or other.

Previous laser processing literature has, with few exceptions, been divided into two groups:

1. Rather complicated, very detailed work for specialists
2. Enthusiastic but uninformative advertising

Personal experience has often left me mentally defeated by the first approach and suspicious of the second. I have tried, therefore, to produce a book which is as readable and informative as possible. Mathematical equations have been reduced to a minimum and the style is generally conversational and descriptive. Most of the chapters begin by expressing, as simply as possible, the basic principles which will be dealt with in more depth later in that chapter. In this way it should be possible for the reader to pick up the basics by a quick read of the first few pages of any section. Chapters 9 and 10 are rather more detailed and scientific in content than the rest of the book and Chap. 11 contains a bibliography of a selection of work on the subject.

The author would appreciate any requests, advice or information from laser users which could make future editions of this type of work more useful.

Acknowledgements

Any technical author owes a large debt to an enormous number of people. I would like to thank all my colleagues and friends in the laser applications industry for their advice and help over the past ten years or so.

Well deserved especial thanks must go to:

Prof. W.M. Steen of Liverpool University, UK

Prof. I.A. Menzies of Loughborough University, UK

Prof. Claes Magnusson of Lulea University, Sweden

Dr. Anders Ivarson of Lulea University, Sweden

Dr. J.N. Kamalu of The University of the West of England, UK

Klaus Frass of Rofin-Sinar

David Price of Electrox

Pierre Scheyvaerts of Rofin Sinar

Mike Sasnett of Coherent Inc.

and the long suffering team at Laser Expertise who have been waiting for me to finish this book so that I can "get on with some real work": Anton Schwarz, Clem Young, Graeme Ellis, David Rudd and others. Last but not least I would like to thank Margaret Lee for all her efforts deciphering my handwriting and producing the typed script.

The book is, of course, dedicated to my mother without whose advice I would now be an unemployed musician. (I hope that after reading the book you don't decide that her advice was something I should have ignored!)

Notes on the Units Used

The units of measurement chosen for this book are intended to render the information as useful as possible to the widest range of readers. For this reason laser cutting speeds are given in metres per minute and simultaneously in inches per minute even though it is generally accepted that time should be measured in seconds in technical literature (m/min and in/min are the units generally used by the cutting industries).

Gas pressures are given in bar and pounds per square inch (psi) as these scales are in more widespread use than the more rigorously scientific unit of the pascal (Pa).

Scientific workers usually have a calculator ready to hand and will be less inconvenienced by having to work out unit conversions than would a "hands on" laser operator. If such conversions are necessary the following list of conversion factors should be of some help.

Distances and Velocities

1 m = 39.4 in (~40)
1 in = 25.4 mm (~25)
1 thou (0.001 in) = 25.4 μm (~25)
1 μm = 0.0394 thou (~0.04)
1 m/min = 16.67 mm/s
1 in/min = 0.423 mm/s
1 m/min = 39.4 in/min (~40)
1 in/min = 0.025 m/min

Pressure

1 bar = 14.5 psi
1 bar = 100 kPa
1 bar = 750 torr
1 psi = 0.0689 bar

1 psi = 6.89 kPa
1 psi = 51.7 torr

Temperature

To convert °C to absolute temperature (K), simply add 273, e.g.

$$25\,^{\circ}\mathrm{C} = 25 + 273 = 298\,\mathrm{K}$$

Chapter 1

The Basic Principles

This chapter deals with four sets of basic principles, all of which are important to the laser user:

1. The cutting process
2. How materials respond to laser light
3. How CO_2 lasers work
4. Laser modes

These topics will be dealt with in a purely conversational and conceptual manner at this stage. A more in-depth approach to these subjects can be found in later chapters.

1.1 The Cutting Process

The basic mechanism of laser cutting is extremely simple and can be summarised as follows:

1. A high intensity beam of infrared light is generated by a laser.
2. This beam is focused onto the surface of the workpiece by means of a lens.
3. The focused beam heats the material and establishes a very localised melt (generally smaller than 0.5 mm diameter) throughout the depth of the sheet.
4. The molten material is ejected from the area by a pressurised gas jet acting coaxially with the laser beam (see Fig. 1.1). (N.B.: With certain materials this gas jet can accelerate the cutting process by doing chemical as well as physical work. For example, steels are generally cut in a jet of pure oxygen. The oxidation process initiated by the laser heating generates its own heat and this greatly adds to the efficiency of the process.)
5. This localised area of material removal is moved across the surface of the sheet thus generating a cut. Movement is achieved by manipulation of the focused laser spot (by CNC mirrors) or by mechanically moving the sheet on a CNC *X*-*Y* table. "Hybrid" systems are also available where the material is moved in one axis and the laser spot moved in the

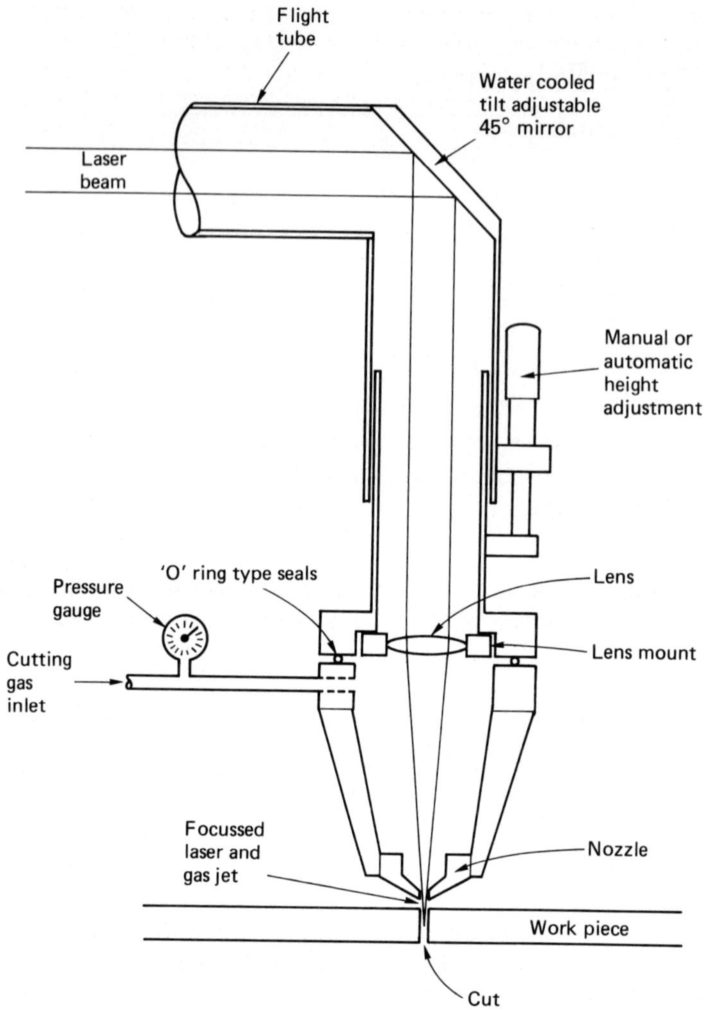

Fig. 1.1. A schematic of laser cutting. The lens mount or the nozzle (or both) can be adjusted from left to right or into and out of the plane of the sketch. This allows for centralisation of the focused beam with the nozzle. The vertical distance between the nozzle and the lens can also be adjusted.

other. (N.B.: Fully robotic systems are now available for profiling three-dimensional shapes although for the sake of simplicity the majority of this book will deal with the cutting of flat sheet on a simple *X-Y* movement system.)

The first industrial use of CO_2 lasers was the cutting of plywood dye boards for the packaging industry. Since this early application in 1971 the technology has developed enormously and lasers are now used with great commercial and technical success to cut almost any material. A cursory review of the literature available will reveal the enormous scope of application, from drilling the holes in baby feeder teats to cutting super alloy panels for the aerospace industry. The laser cutting process has a number of advantages over competing technologies which have ensured the growth of this branch of industry:

1. It is a non-contact process which means that material needs only to be lightly clamped or merely positioned under the beam. Flexible or flimsy materials can be cut with great precision and do not distort during cutting as they would when cut by mechanical methods.
2. The cut width (kerf width) is extremely narrow (typically 0.1 to 1.0 mm) and so very detailed work can be carried out without the restriction of a minimum internal radius imposed by milling machines and similar mechanical methods.
3. The process is fully CNC controlled. This, combined with the lack of necessity for complex jigging arrangements, means that a change of job from cutting component "A" out of steel to cutting component "B" out of a polymer can be carried out in seconds.
4. Although laser cutting is a thermal process, the actual area heated by the laser is very small and most of this heated material is removed during cutting. Thus, the thermal input to the bulk of the material is very low, heat affected zones are minimised and thermal distortion is generally avoided.
5. The process cuts at high speed compared to other profiling methods. For example a typical 1200 W laser will cut 2 mm thick (0.08 in) mild steel at 6 m/min (~240 in/min). The same machine will cut 5 mm (0.2 in) thick acrylic sheet at ~12 m/min (~480 in/min).
6. In most cases (e.g. the two examples given above) the cut components will be ready for service immediately after cutting without any subsequent cleaning operation.
7. Owing to the CNC nature of the process, the narrowness of the kerf width and the lack of mechanical force on the sheet being cut, components can be arranged to "nest" very close together. This reduces wastage of material to a minimum. In some cases this principle can be extended until there is no waste material at all between similar edges of adjacent components. The laser cut separates the two components which therefore "share" the cut line. This is, of course, an extremely effective use of material and laser time but is generally only applicable where two components can be arranged to share a straight line cut.
8. Although the capital cost of a laser cutting machine is substantial, the running costs are generally very low. Many industrial cases exist where a large installation has paid for itself in under a year.
9. The process is extremely quiet compared to competing techniques, a factor which improves the working environment and the efficiency of the operating staff.
10. Laser cutting machines are extremely safe to use in comparison with their mechanical counterparts.

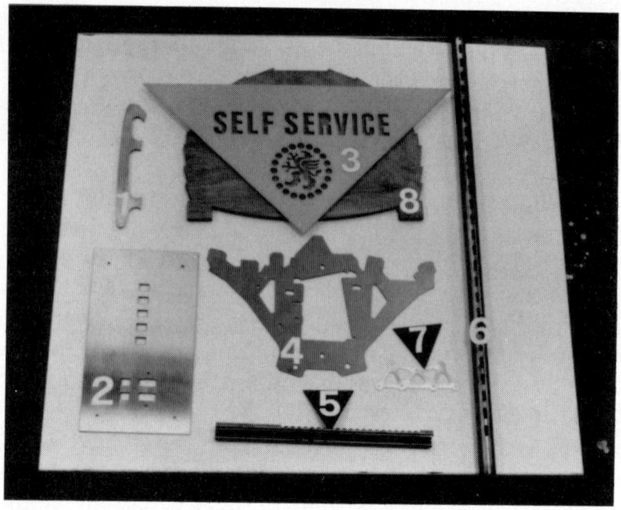

Fig. 1.2. Typical examples of job shop work cut by a 400 W CO_2 laser. 1: An ice skate blade cut out of a high carbon steel blank. 2: A control panel for electronic switchgear cut out of 2 mm thick brushed stainless steel. 3: A display sign cut out of 6 mm thick acrylic sheet. 4: A complex engineering component cut out of 2 mm thick mild steel. 5: A castellated profile in 2 mm thick mild steel, cut for an adjustable position slide. 6: Slots cut in 2 mm thick mild steel tubing of 30 mm diameter. 7: A load distribution plate made from a titanium alloy. 8: A sheet of 12 mm thick marine plywood cut to form one of the components of a "kit" boat.

As an example of the flexibility of the process Fig. 1.2 shows a photograph of a number of typical components cut by a laser cutting job shop using a 400 W laser and a CNC controlled *X-Y* table.

Of the eight items included in Fig. 1.2 five are made from steel (mild and stainless), one is a non-ferrous metal (titanium alloy) and the other two are non-metallic materials. These proportions are a fair reflection of the distribution of applications of CO_2 laser cutting. Although the work shown in Fig. 1.2 was carried out using a c.w. (continuous wave) laser many cutting lasers are also equipped with a facility for producing a pulsed beam. Pulsed cutting can be used to good effect to produce very high quality cuts in metallic materials and is especially useful for a widespread application not shown in Fig. 1.2; the cutting and scribing of ceramic materials.

As indicated by Fig. 1.2, the most common cutting application of CO_2 lasers is the profiling of steels, particularly mild steel. For this reason a large proportion of this book is given over to the subject of cutting this material (see Chaps 2, 6 and 10). Although the main impetus behind the growth of CO_2 laser cutting has been the effectiveness with which these machines can profile steels, the cutting mechanism itself is not yet fully understood. It is clear that the laser acts as a heat source to initiate and accelerate the exothermic oxidation reaction of iron in the oxygen cutting jet, but some fine details of the interaction are still eluding clarification (see Chap. 10). On a more practical level, a great deal of useful information has been generated which has led to the development of the processing techniques which will be discussed later in this text.

Perhaps the second most important group of materials in the subject of laser cutting are non-metals, particularly acrylic sheet, plywood and ceramics. The cutting mechanisms in these cases are rather more straightforward than the case of mild steel and a full description can be put forward alongside the practical techniques which influence the profitability of the process (see Chap. 4).

Non-ferrous metals form a third group of laser cut materials for which the cutting mechanism is readily understood. The laser generates a localised weld pool which is sheared by the incident pressure of the cutting gas jet in order to generate a cut. In some cases an exothermic oxidation reaction can also be employed to accelerate the process. As later discussions will show (see Chap. 3), these alloys can present problems as a result of their high reflectivity to the laser light combined with a high thermal conductivity and an inefficient or inappropriate oxidation reaction. In spite of these problems most non-ferrous alloys can be cut at commercially acceptable rates by CO_2 laser.

1.1.1 Cutting Parameters

Figure 1.1 shows a schematic of a typical cutting head. It is clear that the focusing lens forms part of the seal of a pressure chamber fed from the gas inlet. The only exit from this chamber is the nozzle which therefore emits a pressurised gas jet as well as the focusing beam. Typical nozzle diameters range from 0.8 to 2.0 mm (0.03–0.08 in) depending on the application and personal preference. The subject of nozzle choice and design is far more complex than would be assumed by merely considering it to be a circular opening in the pressure system, and this subject will be dealt with in Chap. 5. As a rule of thumb, a copper nozzle constructed along the lines demonstrated in Fig. 1.3 with a final aperture of between 1.0 and 1.5 mm (0.04–0.06 in) diameter can be used successfully in 90% of applications. Gas pressures (as measured at the inlet to the chamber; see Fig. 1.1) used during

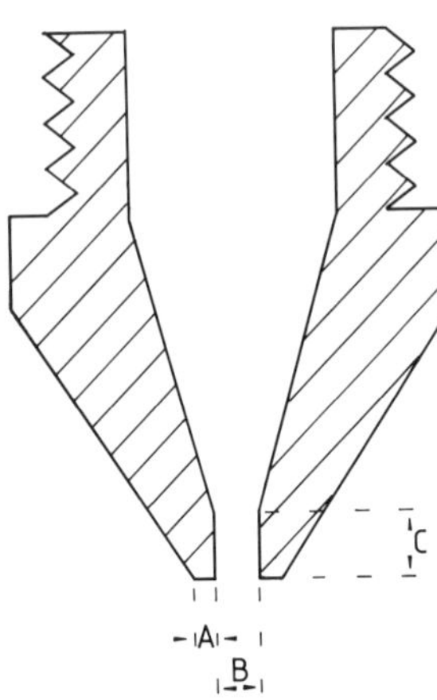

Fig. 1.3. The basic design for a cutting nozzle (cross section). Nozzles are usually made of copper. Approximate dimensions: A, ~1.0 mm (0.04 in); B, 0.8–2.0 mm (0.03–0.08 in); C, 1.0–2.0 mm (0.04–0.08 in).

cutting range from 0.1 to 15 bar (1.5–200 psi) but moderate pressures will suffice for most applications (e.g. 2 bar or 30 psi). Commonly used lens focal lengths are 5 in (127 mm) and 2.5 in (63.5 mm) and these lenses are usually constructed from zinc selenide, although other materials are available. Laser powers used for cutting range from 25 W to 2.5 kW, but continuous improvements in beam output quality are increasing this upper limit.

1.2 How Materials Respond to Laser Light

At first glance a typical laser cutting power of 1 kW does not seem sufficient to cut through metals and plastics. After all, a fan heater for a small room can exceed this power and light bulbs for normal use are often rated at 150 W. The key to the success of laser cutting was hinted at earlier by mention of the extremely localised nature of the molten zone generated during cutting. It is not just the total amount of energy available to the material which creates a melt, it is primarily the density of this energy at the focal point or "hot spot". Laser light can be focused much more effectively than more traditionally generated light as demonstrated by Fig. 1.4. Lenses are devices which can bend incident light to converge in a particular area called the focal point. The word "point" in this case is rather misleading as any lens will bend (or diffract) different colours (or wavelengths) of light by different amounts. The practical result of this is that if a white light source (which is in fact a combination of many colours) is focused by a normal magnifying glass lens, the blue light is diffracted more effectively than the red light and so a range of focal positions is established, as shown in Fig. 1.4. Another feature of traditional light sources which spreads out the so called "focal point" is the diversity of directions the light is travelling in before it passes through the lens. Figure 1.4 shows that a parallel beam is focused to a much smaller point than a non-parallel beam.

The light emitted by a CO_2 laser is monochromatic (i.e. a single colour or wavelength) and parallel in nature and it is these two properties which allow it to be focused down to a small spot with a very high energy density.

A typical focused spot diameter is 0.3 mm which, for 1 kW, gives an average energy density of $1.4 \times 10^{10}\,W/m^2$ or $1.4\,MW/cm^2$ ($0.4\,in^2$), which is enough to boil most materials if they are exposed to the beam for a sufficient length of time. The incident high energy light beam heats by setting up thermal vibrations within the material until its structure is degraded. This degradation can be physical (as in the melting of metals) or chemical (as in the case of burning wood). It is interesting to note that all metals at room temperature are highly reflective to the CO_2 laser light wavelength (10.6 µm). As the metals are heated they become better absorbers and are thus more effectively heated to become even more absorptive and so on. This heating–increased absorption–heating cycle is difficult to set up in materials which are very highly reflective such as copper and aluminium. These non-ferrous metals combine a high reflectivity with a high thermal conductivity and this reduces the efficiency of the cutting process. For copper and aluminium, commercial cutting rates are still

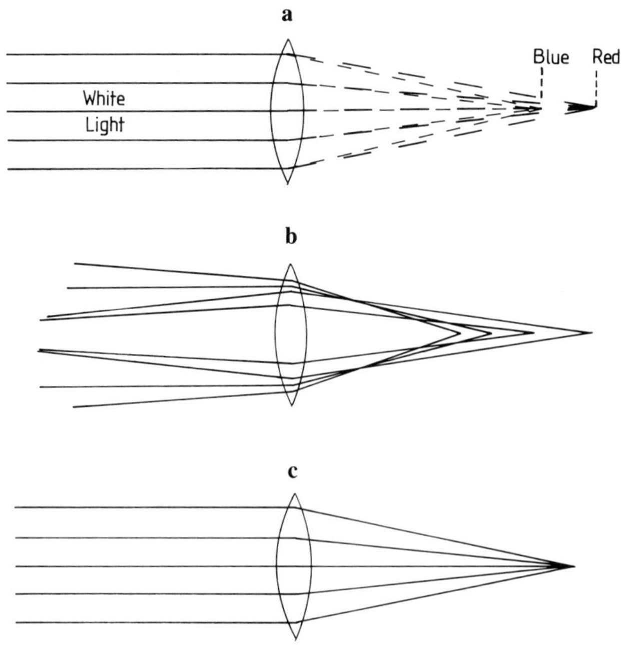

Fig. 1.4. A comparison between the focusing of white light and laser light from other sources. **a** If white light is focused, a range of focal positions is established as each colour focuses at a different distance from the lens. **b**. If non-parallel light is focused, a range of focal positions is established depending on the angles of incidence of the various components of the original beam. **c**. Laser light is monochromatic (single wavelength or colour) and parallel and can therefore be focused to a very intense small diameter spot.

possible (see Chap. 3), but in the case of the most extremely reflective and conductive metal, gold, CO_2 laser cutting is impracticable above foil grades. In contrast to this property of metals, most non-metallic materials are highly absorptive at a wavelength of 10.6 μm and this is often combined with a low thermal conductivity. Most non-metals are therefore cut with high efficiency by CO_2 lasers. The most usual effect of the intense laser light is to produce a localised melt which is then blown away by the gas jet acting co-axially with the beam. In the case of thermoset plastics or wood (i.e. materials which do not melt) the laser heat acts to break the bonds holding the molecules together. The resulting cut is generated by a localised burning process which, in the case of wood, produces much the same carbon based smoke as normal burning (see Chap. 4).

1.3 How CO_2 Lasers Work

Figure 1.5 presents a schematic of a CO_2 laser. The principles of the lasing action are identical for any CO_2 laser regardless of the design. The major

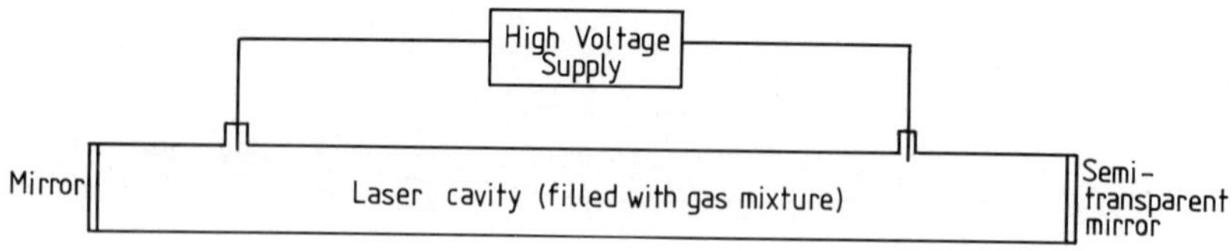

Fig. 1.5. A schematic of a CO_2 laser.

design differences between laser types are concerned mainly with ways of exciting and cooling the gas mixture in the laser cavity.

In any CO_2 laser a high energy stream of electrons is passed across a specific low pressure gas mixture consisting of CO_2, nitrogen and helium, typically in the proportions 1:5:20 by volume, respectively. The typical total pressure of the mixture is 20 torr or 27 mbar. Nitrogen and helium are added to the CO_2 to render the lasing action more efficient in ways which will be described shortly.

The basic idea is that the input to the laser is mechanical energy over a very broad spectrum in the form of the rapidly moving electrons. These electrons pass energy on to the gas molecules by collision and by the following mechanism the CO_2 generates a pure, parallel beam of infra-red light.

1.3.1 The Storage and Release of Energy from a CO_2 Molecule

It has been already mentioned that the nitrogen and helium are included to improve the efficiency of the laser. The actual lasing reaction involves only CO_2 and so this can be described in isolation. (A pure CO_2 laser does work but is only ~1% efficient, compared with the 10%–25% efficiency of the mixed gas laser.)

Carbon dioxide is a triatomic (three-atom) molecule with the two oxygen atoms attached to either side of the carbon. A CO_2 molecule can store energy by becoming distorted in the same way as a spring can. However, molecules are governed by the laws of quantum mechanics which say that only certain specific amounts of energy can be stored by each type of molecule. The analogy of a spring which can be compressed to any position is therefore misleading. A better model is described by Fig. 1.6.

This sprung loaded ratchet type mechanism models the CO_2 molecule rather well. It shows that only certain values of energy can be stored and that the release of energy would involve the movement from one ratchet tooth to the next. The positions of these "ratchet teeth" are usually described by an energy spectrum as shown in Fig. 1.10. For the action of lasing a material should have two energy states rather far apart, the top one of which is neither too stable nor too liable to spontaneous release. These two energy states for CO_2 are shown schematically in Fig. 1.6. The reason for the electric discharge in a CO_2 laser is to bombard the CO_2 molecules with high speed projectiles (the electrons) in order to reach the upper of these two states by "cocking" the ratchet to the correct level. The molecules which achieve this upper energy level can be referred to as "excited". After

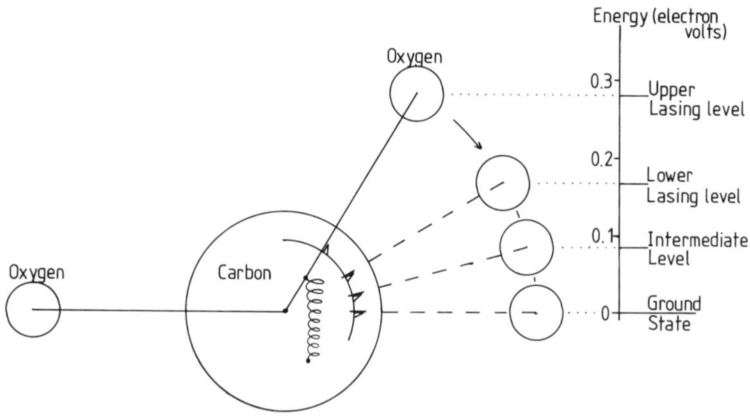

Fig. 1.6. The carbon dioxide molecule depicted as a sprung loaded ratchet.

only a few microseconds of bombardment by the electric discharge a large proportion of the CO_2 molecules will be in the "excited" condition and this is the condition under which lasing can be achieved.

1.3.2 Lasing

First of all a few of the "excited" molecules spontaneously relax. The result of any of these relaxations is the release of a packet of energy in the form of a "particle of light" or photon, as shown in Fig. 1.7. The energies of these photons are identical to the drop of mechanical energy the molecule experiences by relaxing. As the molecules all relax from the same "excited" state, the photons are all of the same energy. The energy of the photons determines their wavelength (or colour) and so they must all be of the same wavelength.

The actual lasing reaction occurs after one of these spontaneously emitted photons collides with another "excited" molecule. It was mentioned earlier that the "excited" state of the molecule should neither be too stable nor too unstable. A small amount of stability is necessary in order to prevent the molecules all relaxing before they can be involved in a collision with a photon. The instability is responsible for what happens next; the "excited" molecule cannot accept the incoming energy as it does not fit its own energy spectrum but it now finds itself disturbed to the extent of not only rejecting

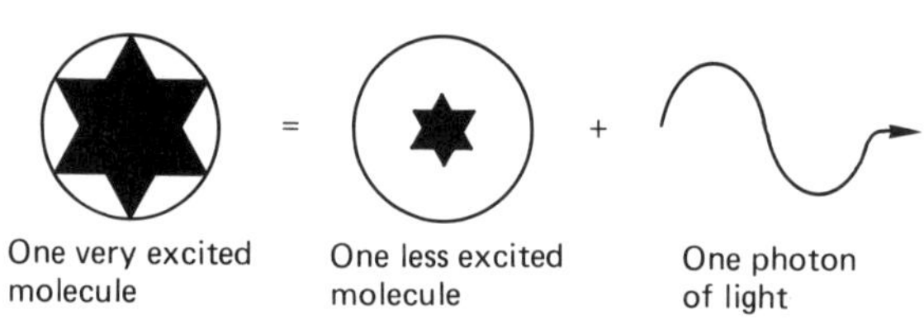

Fig. 1.7. The relaxation of a CO_2 molecule causes the generation and emission of a photon.

the incoming photon but of releasing its own energy as a second photon. This second photon leaves the molecule at the same time as the first, is of exactly the same energy (and therefore wavelength) and travels in the same direction. It is therefore a duplicate of the first photon. These two photons go on to collide with other "excited" CO_2 molecules to generate more and more copies of themselves as an avalanche effect. This avalanche effect or Light Amplication (LAser) can be seen happening in a number of directions in Fig. 1.8 depending on direction of the first spontaneously relaxed photon.

Although this reaction of

1 photon + 1 "excited" CO_2 molecule
= 2 photons + 1 relaxed CO_2 molecule

is the lasing mechanism, it does not constitute a laser. The laser as a machine only begins to work when one or more of these initial photons is produced travelling along the line perpendicular to the mirrors shown in Fig. 1.9. In this case the eventual avalanche of photons is not lost by absorption by the walls of the cavity enclosing the gas but travels up and down between the two mirrors until the maximum number of "excited" molecules are continuously being Stimulated to Emit Radiation (laSER) (the Radiation being of course the photons of light). Once a molecule has released its photon it is very quickly re-excited by the high energy electrons and so the cycle continues. One of the two mirrors shown in Fig. 1.9 is only semireflective in order to release approximately half of the light generated for subsequent industrial use. The other half of the light is reflected back into the laser cavity to ensure the wholesale multiplication of photons by the lasing process. In this way a continuous beam of laser light is emitted from the half mirror or output window (also known as the output coupler). If a pulsed output is required this can be achieved by electronically switching the electric discharge on and off.

1.3.3 The Role of Nitrogen

The efficiency of a CO_2 laser is determined by the effectiveness of the system in exciting the relaxed CO_2 molecules up to the required upper energy level. The efficiency of doing this simply by bombardment with high velocity electrons is very low but fortunately nitrogen can be used as a catalyst. Nitrogen is a diatomic (two-atom) molecule and therefore has a much less complicated energy spectrum than the triatomic CO_2 because it can store energy in far fewer ways. If nitrogen is bombarded with the correct level of electric discharge it is readily excited to an energy level very close

Fig. 1.8. The avalanche effect of light amplification takes place as photons collide with "excited" CO_2 molecules which then give up their stored energy in the form of further (identical) photons. Stage 1. Spontaneous emission of photons from excited CO_2 molecules which then relax to a less excited stage (see Fig. 1.7). Stage 2. Spontaneously emitted photons collide with excited molecules in their path which then relax to release a further photon; this process is "*S*timulated *E*mission" (LA*SE*R). Stage 3. After a number of photon–molecule collisions an avalanche effect is built up. Each avalanche follows the direction of its original, spontaneously released photon.

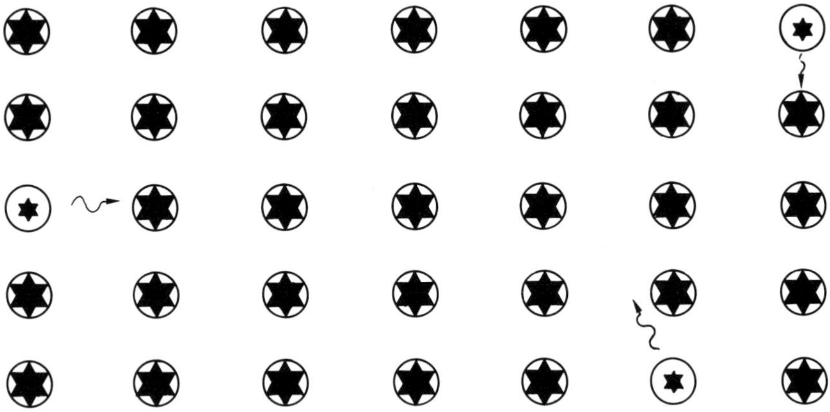

Stage 1.

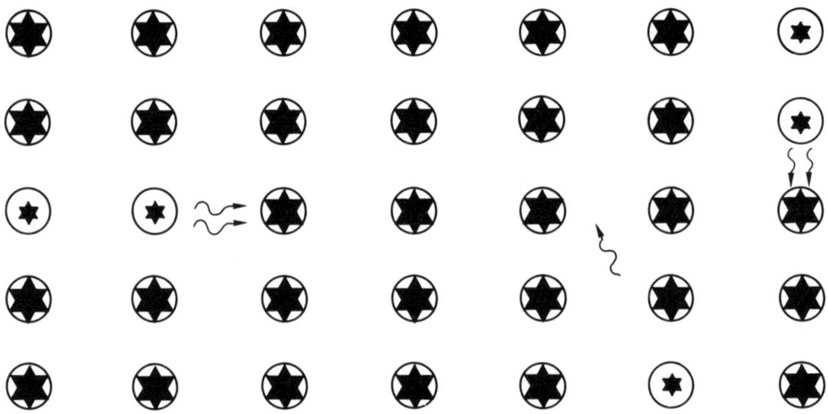

Stage 2.

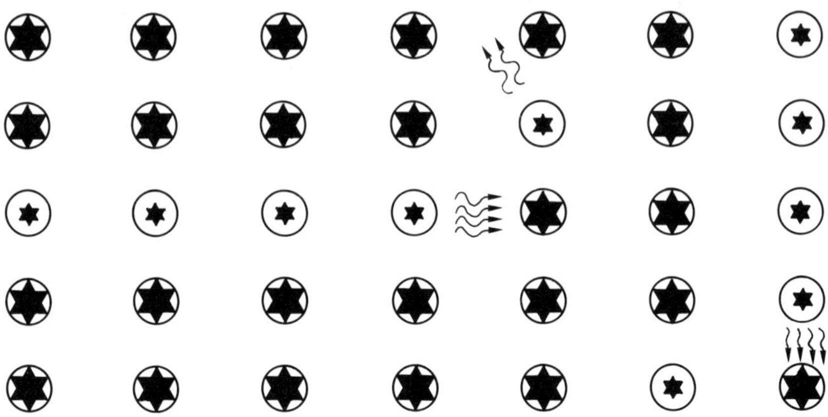

Stage 3.

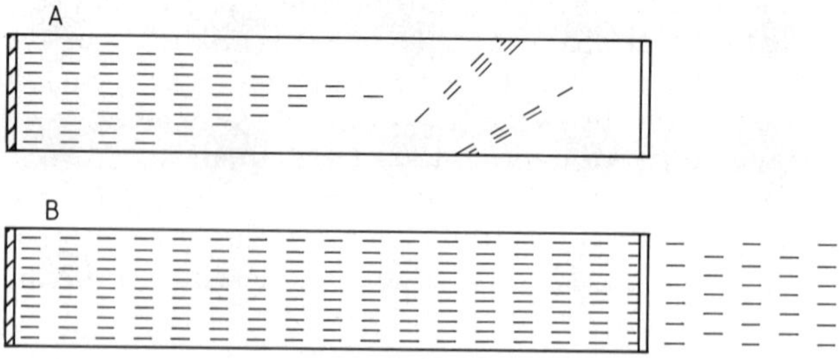

Fig. 1.9. As the avalanche effect becomes established along the axis of the laser cavity, the light reflects off the mirrors within the cavity and the lasing process becomes self sustaining. A proportion of the light thus generated is continuously "leaked" out of the half mirror to be used for materials processing.

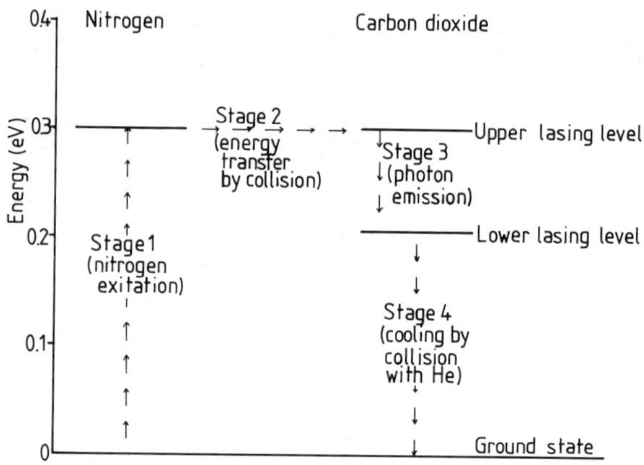

Fig. 1.10. The energy spectra of CO_2 and nitrogen.

(within 1%) to the upper energy level required of CO_2 for lasing (see Fig. 1.10). The nitrogen itself does not "lase" effectively because this energy level is rather too stable. (Nitrogen lasers do exist but are not relevant to our discussion.) The nitrogen will, however, pass on this energy to CO_2 molecules during any collision. In this way nitrogen "packages" the energy present in the electron stream and transfers it to the CO_2 to result in a large population of "excited" CO_2 molecules.

1.3.4 The Role of Helium

Throughout the previous discussion the word "relaxed" has been used to describe the state of a CO_2 molecule once it has released its stored photon. "Relaxed" implies that the CO_2 has returned to its initial state (i.e. the ground state), but this is not the case. Referring back to our ratchet model, the release from a highly excited state takes us only to the next ratchet step downwards. This lower energy level is not far removed from the ground state, but all the energy must be removed from the CO_2 in order for it to be able to accept the package of energy offered by the nitrogen (a partially energised CO_2 molecule cannot accept the nitrogen "packaged" energy because the sum of the two energies would not fit the energy spectrum of CO_2). The complete de-excitation of the CO_2 molecules down to the ground state is very inefficient in a CO_2–N_2 laser, but the addition of helium to the gas mixture helps matters tremendously. The partially energised CO_2 molecules lose their stored energy by colliding with the helium atoms. In this way the stored energy is transformed into kinetic energy.

1.3.5 Gas Mixture Cooling

The final step in the system of energy exchanges which make up a CO_2 laser is to cool the whole gas mixture, because once the helium has absorbed the last vestiges of stored energy from the CO_2 it begins to warm up. The energy spectra which govern the working of the laser are disturbed by changes in temperature and so it is necessary to keep the whole gas mixture at a moderate temperature. There are two basic methods of cooling the gas mixture which have lent their names to the two most commonly available CO_2 laser types: slow flow and fast flow. The major features of these laser types will be dealt with in the following section on laser modes. Briefly, in slow flow machines the walls of the lasing cavity are kept at approximately room temperature by oil or water cooling and the heated gas contained within the cavity loses heat by conduction to these walls (see Fig. 1.11). Fast flow designs incorporate a refrigeration unit outside the laser cavity through which the gas is continuously pumped (see Fig. 1.12).

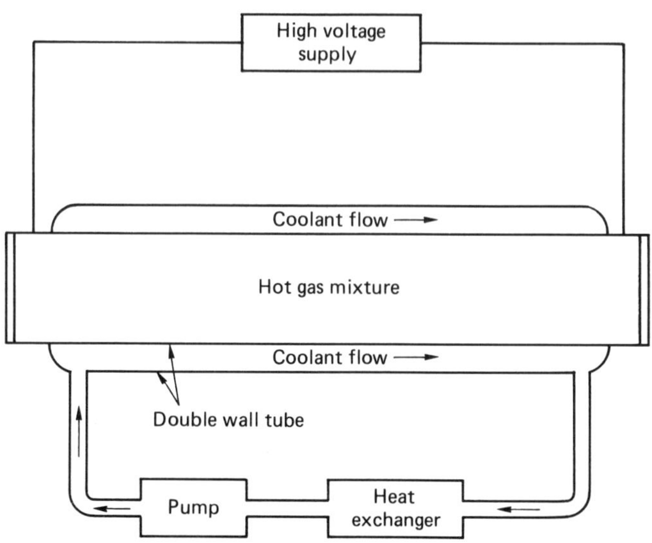

Fig. 1.11. A schematic of the gas cooling mechanism of a slow flow laser.

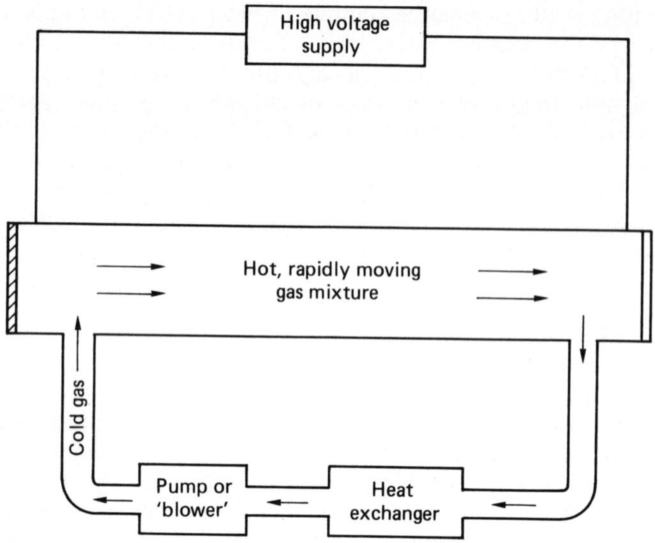

Fig. 1.12. A schematic of the gas cooling mechanism of a fast flow laser.

1.3.6 Gas Recirculation

In all types of CO_2 laser a certain amount of new gas is continuously added to the chamber and the same volume is exhausted, usually through the vacuum pump. The reasons for this renewal of the gas content of the laser is the occasional break up of the gas molecules by the electron stream. This break up "introduces" new gases to the mixture (e.g. CO, NO_2 etc.) which interfere with the lasing action and render the machine less efficient.

The gases are therefore a major running cost and to minimise wastage a recirculation system is introduced to fast and slow flow machines. This recirculation involves the passage of a large proportion of what would be exhausted gas through a series of catalytic converters to recombine the atoms into the required molecules. This "repaired" gas is then reintroduced to the general mixture in the laser (see Figs 1.11 and 1.12).

1.3.7 A Summary of the Basic Energy Exchanges in a CO_2 Laser

Two summaries are given here, the first of which describes the main interactions including the initiation of the laser beam. A second, much shorter summary explains continuous wave operation which assumes the existence of an avalanche of lasing reaction generated photons along the axis of the laser cavity. Secondary energetic interactions such as the direct excitation of CO_2 by the electron stream or dissociation of molecules have been ignored.

1.3.7.1 Including the Lasing Beam Initiation

1. High speed electrons collide with nitrogen molecules which are thereby excited to a particular energy level very close to the upper lasing energy level of CO_2.

2. Energised nitrogen molecules collide with CO_2 passing on this "package" of energy.
3. A few "excited" CO_2 molecules spontaneously relax almost all the way down to their ground state. As a result of relaxation from the upper lasing energy level to the lower, a photon is emitted.
4. The spontaneously emitted photons travel through the gas mixture in the laser cavity until they collide with another "excited" CO_2 molecule. This collision upsets the semistable state of the "excited" molecule which therefore rejects the incoming photon at the same time releasing its own photon which travels off as a twin of the rejected photon.

   1 photon + 1 "excited" molecule
   = 2 photons + 1 partially relaxed molecule

5. These two photons travel in the same direction through the gas experiencing more collisions and therefore generating more "twins" in a type of avalanche effect. These "twins" have the same energy and are therefore of the same wavelength. They also travel in the same direction as the original colliding photon.
6. The direction of the avalanche effect is determined by the original direction of the spontaneously emitted photons. Most of these spontaneous photons are emitted at an angle to the axis of the laser machine and will be lost by absorption by the walls of the laser cavity. A small proportion, however, will be emitted along the axis of the laser and will therefore have a much longer life in the laser cavity involving many more lasing collisions. A dominant avalanche of lasing reactions is thus rapidly established along the axis of the laser. This generates a self-supporting number of identical photons even when half their number are continuously "leaked" out of the half mirror at one end of the cavity.
7. The partially "excited" CO_2 molecules are cooled completely down to their ground state by collision with helium atoms. They are then re-energised by collision with nitrogen molecules which are continuously re-energised by bombardment by the high velocity electrons.
8. The whole gas mixture is cooled by either conduction to the oil cooled walls of the laser cavity (slow flow), or rapid continuous pumping of the gas out of the lasing cavity, through heat exchangers and back into the lasing cavity (fast flow). If this cooling was not carried out the helium would become progressively heated by the partially energised CO_2 molecules and would, eventually, be unable to cool them down to their ground state.

1.3.7.2 During Continuous Laser Operation

1. High speed electrons collide with and energise nitrogen molecules.
2. Energised nitrogen molecules collide with and energise CO_2 molecules (nitrogen is then re-energised by electrons).
3. Previously generated laser photons collide with energised CO_2 molecules which then release most of their stored energy as identical photons.

4. CO_2 is cooled to its ground state by collision with helium and then repeats steps 2, 3 and 4.
5. The whole gas mixture is cooled by conduction or continuous pumping through heat exchangers.

1.4 Laser modes

1.4.1 Mode Types

The "mode" describes the energy density cross section of the laser beam which can be easily (but only approximately) determined by exposing a sheet of acrylic to the unfocused beam for a few seconds. As the beam irradiates the material the acrylic is rapidly evaporated at rates determined by the local energy density. The laser beam therefore leaves an imprint related to its distribution of energy in the plastic; an example of such a "beam print" is shown in Fig. 1.13. Four basic types of mode are available from CO_2 lasers:

Gaussian mode. This mode has a cross section which is a single dense circle of energy which increases towards the beam centre at the same rate as a Gaussian curve (see Fig. 1.14).

Doughnut mode. This unscientific but descriptive name is given to a mode which is circular in cross section but is hollow in the centre (see Fig. 1.15).

Multimode. This name covers modes which have an energy distribution which is a combination of Gaussian, doughnut and other more complex (or "higher order") modes (see Fig. 1.16). In some cases a laser designed to generate a Gaussian mode will produce a multimode if poor quality or

Fig. 1.13. A typical "mode print" or "mode burn" in an acrylic block. This beam (which irradiated the plastic from the bottom of the picture) shows good axisymmetry and a close to Gaussian profile (see Fig. 1.14).

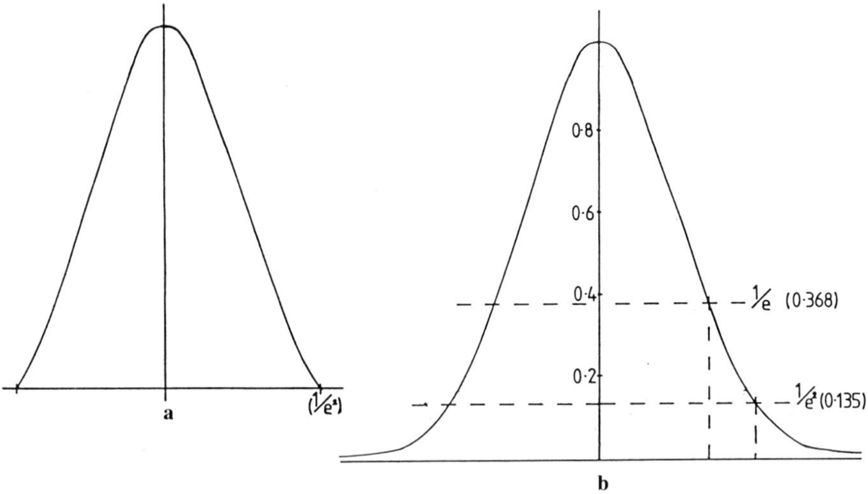

Fig. 1.14. **a** Any cross section of a "Gaussian" beam should have an energy distribution which follows this type of curve (see Fig. 1.13). **b** The curve in **a** is derived from the mathematically calculated Gaussian curve. As this curve continues decreasing to infinity the laser industry has found it necessary to use a truncated form of the curve which is usually cut off at the $1/e^2$ level as shown. Some scientists prefer to truncate the curve at the $1/e$ level (for further information on modes see Chap. 9).

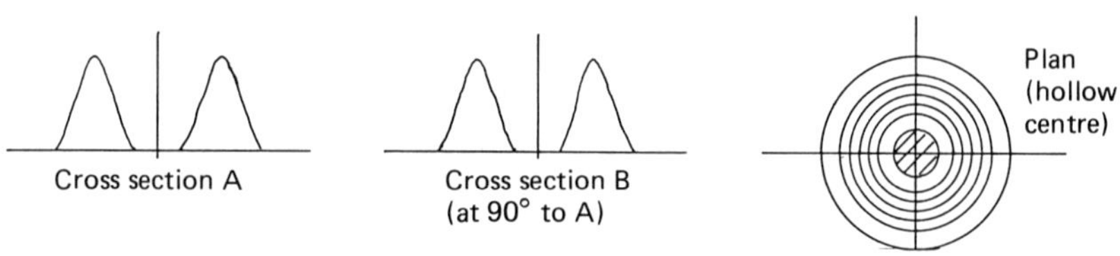

Fig. 1.15. A doughnut mode (TEM_{01^*}).

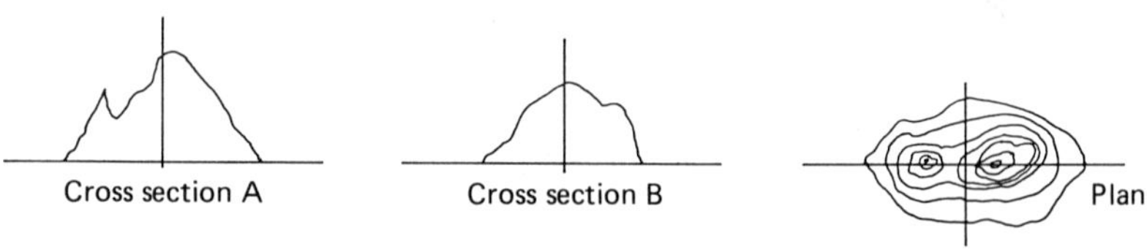

Fig. 1.16. A multimode generated by the superimposition of other modes and, in some cases, imperfect optical components.

dirty optical components are used. Other laser operating parameters (e.g. gas mix quality) can also degrade a Gaussian beam to multimode.

Single high order modes. It is possible in CO_2 as well as all other lasers to adjust the internal mirrors so that they do not lie exactly perpendicular to the lasing axis (i.e. the mirrors do not quite face each other). The outcome of this condition is that destructive interference of the laser light with itself occurs in various areas across the surface of the mirrors. This in

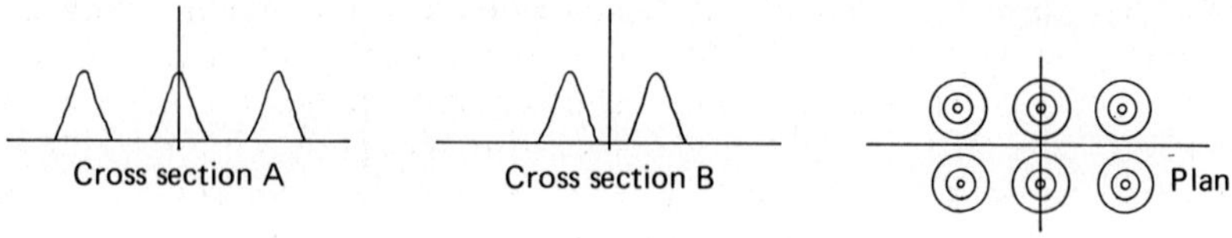

Fig. 1.17. A high order mode (TEM_{21}).

turn leads to distinct "gaps" or areas of low energy within the output beam cross section. A slight tilt of one mirror in one axis can, for example, generate an output beam which consists of two or more parallel strips of energy. Tilting a mirror in both axes can produce a complex pattern of dots of energy. Modes of this type (e.g. see Fig. 1.17) which suffer from destructive interference within the lasing cavity are less powerful than low order modes such as Gaussian or doughnut as they are obviously generated with less efficiency. They are therefore generally to be avoided by the CO_2 laser materials processor. Another disadvantage is the inferior focusing characteristics of this type of mode.

1.4.2 A Quick Guide to TEM Nomenclature

Throughout the scientific and commercial literature CO_2 laser modes have their name accompanied by a TEM_{xy} number, for example Gaussian (TEM_{00}). TEM stands for Transverse Electromagnetic Mode and the xy value describes the number of gaps or areas of low energy density within the cross section of the beam. Acrylic sheets can be used to take a print of the beam, which might, for example, have a number of isolated spots of energy as shown in Fig. 1.17. To find the TEM_{xy} number, count the gaps within the beam print, first from left to right (the x value) then from top to bottom (the y value). The case in Fig. 1.17 obviously gives a value of TEM_{21} which can be referred to as a high order mode (only one gap or less is low order).

The single dense spot of a Gaussian beam contains no gaps and therefore is referred to as TEM_{00}. The multimode type beam may have many local variations of energy density (see Fig. 1.16) and can be generated by the superimposition of pure modes or by the corruption of a single mode. Doughnut beams are obviously a special type of TEM_{01} or TEM_{11} as there are no separated areas of energy, but there is a gap in the centre. The nomenclature has therefore been adjusted to TEM_{01^*} where the asterisk implies that the TEM_{01} mode has been spun around its central axis to complete a circle. For a more formal discussion of TEM modes refer to Chap. 9.

1.4.3 The Effect of Laser Design on Mode and the Implications for Material Processing

All CO_2 lasers consist of a chamber within which high energy electrons are passed across a gas mixture of CO_2, nitrogen and helium. Two "end" mirrors determine the axis of the lasing reaction although in many cases intermediate mirrors are used to elongate the optical path or to reduce the overall length of the laser machine. The fundamental mode of most commercially available CO_2 materials processing lasers is either TEM_{00} or TEM_{01^*} (Gaussian or doughnut). This basic mode is primarily determined by the optical arrangement and design of the two end mirrors in the lasing cavity although the purity of the mode can be affected by other design features which will be discussed later.

TEM_{01^*} modes, or doughnut modes, are generated by the type of optical design described in Fig. 1.18 which is known as an unstable resonator. The lasing cavity lies between the two mirrors and the system is designed to "leak" light past the smaller of the two, naturally generating a beam with a hollow centre. Advantages of this design include compactness for the amount of lasing gas involved and robust construction. The large volume of gas involved in the lasing reaction means that high output powers (5–25 kW) are possible but the doughnut mode generated does not focus down to the same extent as the TEM_{00} type beams. This reduction in focusing means that the energy density at the focused position of a 5 kW doughnut mode could be exceeded in some cases by a TEM_{00} mode with a power as low as 1 kW. For this reason the TEM_{01^*} type machines have found their market in the areas where total output power is more important than the smallest

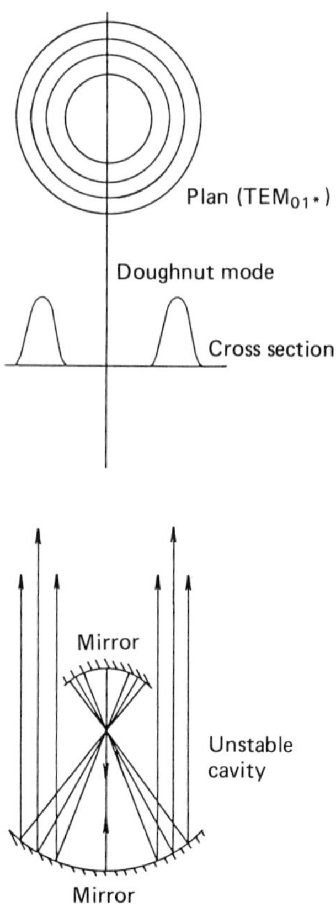

Fig. 1.18. An unstable cavity generating a doughnut TEM_{01^*} beam. The area between the two mirrors is the lasing cavity with laser light experiencing multiple reflections between them (not drawn) before "missing" the smaller of the two mirrors and leaving the cavity. The expression "unstable" refers to the fact that each photon has only a limited lifetime within the cavity and will eventually achieve an output trajectory (cf. stable cavity Fig. 1.19). Electrical discharge across the gas mixture will be in a perpendicular direction to the lasing axis, i.e. from left to right (along the line of the words "unstable cavity"). Other designs of unstable cavity exist, but the principle of operation remains similar to that shown here.

focused spot size. Heat treatment (where the beam is defocused in any case) and welding (fit up tolerances can be relaxed as the focused spot size is increased) are the major outlet for these machines. On the other hand, TEM_{00} Gaussian modes focus down to spot sizes of the order of 0.25 mm diameter and achieve very high energy densities. They are ideal for cutting because they generate the maximum amount of material damage over the minimum area, for this reason they are sometimes described as a "sharp tool" as compared with the doughnut "blunt tool". It should be borne in mind, however, that even the "blunt tool" doughnut beam rarely exceeds 1 mm diameter at the focal position and that in some applications they have been used successfully for cutting.

TEM_{00} modes are generated by the optical configuration described as a "stable cavity"; an example of such a stable cavity is shown schematically in Fig. 1.19. The two end mirrors, of the same overall dimensions, face each other across the long, narrow chamber in which the gas is electrically excited. The usable output is released from the cavity continuously through one of the mirrors which is only half reflective. One (or both) of the mirrors is slightly spherically concave on the inner face and this tends to channel the light oscillating between them towards the centre of the beam. In the best designs this will generate a Gaussian output beam but in any case a single

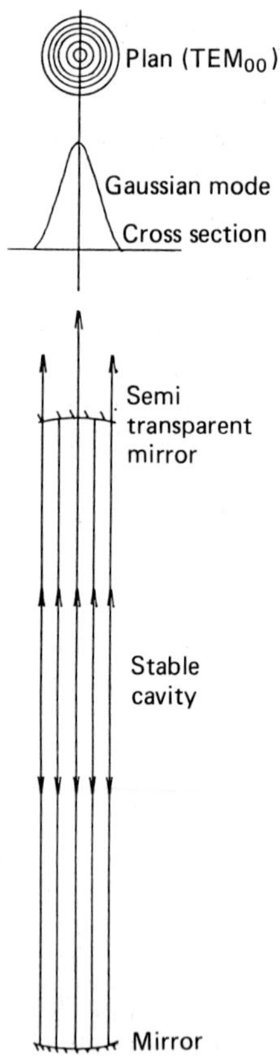

Fig. 1.19. A stable cavity generating a Gaussian TEM_{00} mode. The term "stable" implies the statistical chance that a particular photon may remain in the cavity for an indefinite amount of time if it happens to be repeatedly reflected by the semi transparent mirror. The curvature of the mirrors tends to concentrate the beam towards the centre after multiple reflections (this effect has not been drawn in the interests of simplicity). Electrical discharge in this case is along the lasing direction usually using hollow cylindrical electrodes near each mirror. (These hollow cylinders allow the free passage of the light in the cavity).

dense spot of energy (TEM_{00}) will be the result. Factors which can corrupt the Gaussian nature of the beam include; vibration, poor optical component quality and local variations in pressure of the gas within the cavity.

The local pressure variations are only appreciable in fast flow machines where shock waves etc. may be set up. These can lead to refraction of the beam and degradation of the mode.

One further source of mode corruption is thermal gradients within the plasma discharge in the lasing cavity. The plasma density and temperature are at a maximum towards the centre of the discharge cross section and the thermal gradients are naturally associated with refractive index gradients which affect the mode.

The terms fast flow and slow flow denote the two major groups of CO_2 lasers designed to give a Gaussian output from a stable cavity. The terms refer to the velocity of the gas within the lasing cavity which is determined by the cooling method chosen for the gas:

Slow flow. As mentioned earlier, within a slow flow CO_2 laser the surplus heat generated within the gas mixture as a result of the bombardment by electrons and the inefficient nature (10%–25%) of the lasing process is removed by conduction to the walls of the lasing cavity. One common design is to make the lasing cavity walls out of a glass tube which is itself encased in a coaxial glass tube. The gap between these tubes is filled with a circulating cooled oil which takes the heat away from the inner tube (see Fig. 1.11). Gas velocities within the lasing chamber are low and local variations in gas pressure and refractive index are minimised.

Fast flow. Within a fast flow machine the heated laser gas mixture is continuously pumped away from the lasing cavity, passed through a heat exchanger and then reintroduced to the cavity. Gas velocities can exceed 200 m/s and the gas transport system includes a complex series of bends which can result in local pressure variations (see Fig. 1.12). These fast flow lasers also incorporate high power pumps (roots blowers) which increase the amount of vibration in the system, another source of mode corruption. One great advantage of the fast flow system is the increase in power per unit volume of laser cavity. This cooling system is much more efficient than the rather tranquil conduction–convection mechanism which governs slow flow machines and this has enabled designers to achieve much higher powers from smaller laser machines.

For these reasons of relative size, the majority of slow flow lasers are rated at less than 1.5 kW whereas fast flow machines are available in powers up to tens of kilowatts. The highest powered machines have generally poor quality TEM_{00} modes but, due to recent design improvements, fast flow lasers now exist which give a reasonable approximation to a Gaussian output at powers up to 3 kW.

To summarise these points:

1. TEM_{01^*} modes are not generally suited to cutting as they do not focus down to the small high energy density spot possible from Gaussian modes.
2. The axial symmetry and Gaussian purity of a TEM_{00} mode can be affected by laser design. Slow flow machines generate superior quality

beams to their fast flow counterparts although the machines themselves tend to be substantially larger. Continuous improvements to the design of fast flow machines are showing dividends in the form of improved mode quality.

The effect of mode purity on the cutting process will be discussed in more detail in Chap. 2 (Sect. 2.1.7.1).

Chapter 2

Laser Cutting Steels

The laser cutting industry is dominated by the profiling of steel sheet of both mild and stainless grades. The cutting process is achieved by a combination of laser heating and the chemical reaction of iron with oxygen. This oxidation reaction is exothermic (heat generating) and acts as a secondary energy source which helps to accelerate the cutting process. Basically, the steel is locally heated by the focused laser beam to a temperature at which the iron will ignite in the oxygen jet acting coaxially with the beam. Stainless steel is cut slightly less effectively than mild steel because the presence of chromium oxide in the cut zone tends to seal off the melt from the incident oxygen jet. For this reason the mechanics of cutting are different for mild and stainless steels and it is convenient to discuss them separately.

2.1 Mild Steels

2.1.1 Introduction

It was mentioned in Chap. 1 that during laser cutting materials are subjected to physical and/or chemical damage over a narrow area throughout the depth of the sheet. Mild steel is the most important of the materials which undergo a laser activated chemical reaction during cutting. This chemical reaction (the oxidisation of iron) increases the cutting speed and quality compared with simple laser melt cutting, but has the unfortunate side effect of increasing the sensitivity of the process to certain parameters, particularly

1. Nozzle–beam misalignment or nozzle contamination
2. Poor axial symmetry of the laser mode
3. Contamination of the oxygen supply
4. Local overheating of the workpiece as a result of poor CNC programming

It is for this reason that the troubleshooting section of this book (Chap. 6) is dominated by references to mild steel.

Apart from these four parameters, the cutting process has a remarkably large "operating window" within which excellent cutting results can be achieved, unless maximum possible cutting speeds are required. Most experienced laser users establish the maximum cutting speed for a particular steel sheet thickness and then work at between 80 and 90% of that value.

This practice ensures that the cutting machine is working well inside its limits and is therefore rendered relatively insensitive to minor changes in other process parameters.

A rough guide to the flexibility of the main process parameters when cutting at, say, 85% of the maximum cutting speed is

Laser power. This can be increased by up to 30% or decreased by 10% without seriously affecting the process.

Focus position. Although the focus of the beam should lie on the material surface, defocusing in either direction by ~1% of the focal length of the lens will make only a marginal difference to the cutting process, e.g. for a 100 mm lens the focal point may be positioned ~1 mm above or below the sheet surface. This flexibility allows for good cutting even on non-flat sheet steel. (This subject will be dealt with in Sects 2.1.3 and 10.6.)

Sheet condition. Local areas of moderate surface corrosion or small (±10%) variations of thickness can be accommodated. Also, mechanical damage (dents, scratches etc.) has little or no effect.

Oxygen supply. If the oxygen flow to the nozzle was originally set at an optimum, variations of ±10% can be accommodated. The process is, however, exceptionally sensitive to corruption of the purity of the oxygen supply. An addition of even 1% of contaminant can reduce cutting speeds by up to 30%.

Cutting speed. The cutting speed can be set anywhere between the maximum cutting speed (if all other parameters are optimised) and 50% of that speed. At the ends of this range the material may locally overheat which will give rise to poor cut quality (burned corners etc.; see Sect. 2.1.3).

From these comments it is clear that laser cutting of mild steels is a reliable and rugged process if the four sensitive parameters (the condition and alignment of the nozzle, the symmetry of the mode, the oxygen purity and the avoidance of overheating) are given sufficient attention.

These parameters will be dealt with in detail in Sect. 2.1.7, but it is now appropriate to explain the basic physical and chemical mechanisms which are responsible for the cutting process.

2.1.2 The Cutting Mechanism for Mild Steels

Observation of any mild steel cut edge produced by a combination of a CO_2 laser and an oxygen jet will reveal regular spaced ripples or "striations" along the cut edge as shown in Fig. 2.1. In thin section steel (<4 mm) these striations may be clear and regular from the top of the cut edge to the bottom. However, on thicker sections, the striations may be clear towards the top of the edge but are replaced towards the bottom by more random ripples associated with the flow of liquid out of the cut zone. The striations are evidence that even though the energy input to the cutting zone (i.e. laser energy and oxygen flow) is constant, the interaction between the metal and the incident energy behaves in a cyclic manner. One of the earliest and clearest explanations of this cyclic reaction was developed by a team from Osaka University led by Arata and Miyamoto (Ref. 15, Chap. 11). A

Fig. 2.1. A typical mild steel cut edge showing the distinctive striation pattern. (The sample shown is 2.0 mm thick.)

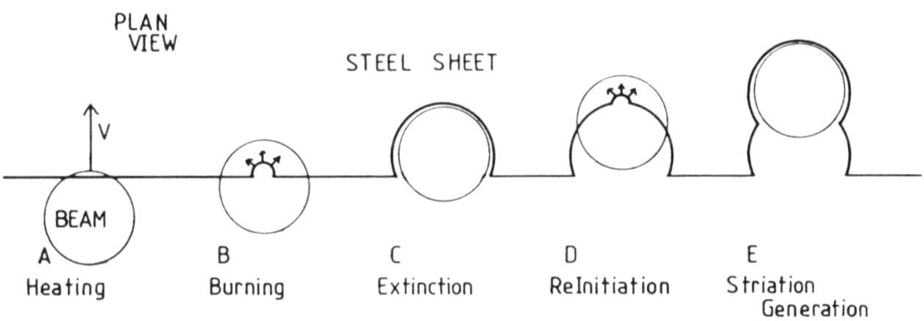

Fig. 2.2. A schematic showing how striations can be generated. (From the work of Arata, Miyamoto et al., Ref. 15, Chap. 11.)

schematic of this simple model is given in Fig. 2.2 and can be described as follows.

1. The laser and its co-axial oxygen jet move onto the edge of the steel sheet and the steel is locally heated by the focused laser to a temperature at which it will ignite in the oxygen stream.
2. The burning front thus established is fed energy directly from the laser and also from the exothermic oxidation reaction. This highly energetic environment causes the burn front to move rapidly away from the centre of the laser beam.
3. Once the laser beam has been left behind, the burning front cools and extinguishes.
4. The moving laser then initiates the re-ignition of the next area to repeat the process and thus generate a pattern of striations along the cut edge.

The cyclic nature of the laser–oxygen–steel reaction which generates the striations is the major clue to a full understanding of the process and many conflicting models have been put forward to explain this phenomenon. The model outlined above provides a basic introduction to the cutting mech-

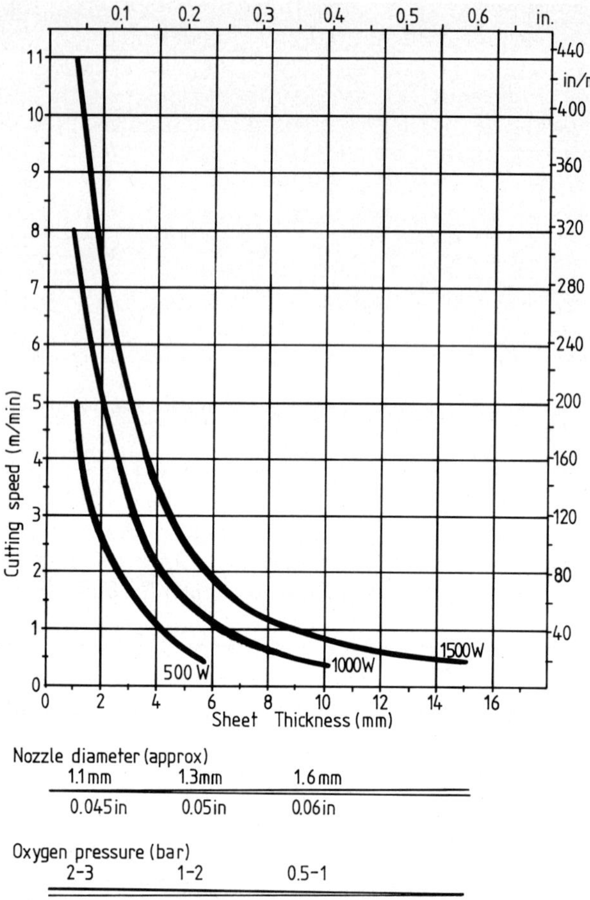

Fig. 2.3. Cutting speeds for mild steel for a number of laser powers. (Guidelines concerning nozzle diameters and oxygen pressures for each thickness are also given.)

anism without resorting to complex mathematics. The interactions in the cut zone are, of course, more complicated than outlined here and involve fluid and gas dynamics as well as thermodynamics (see Chap. 10).

2.1.3 Cutting Speeds and Cut Quality

2.1.3.1 Typical Cutting Speeds

Figure 2.3 shows the cutting speeds possible for mild steel over a range of thicknesses and laser powers. One approximate rule to remember is:

1.5 kW of laser power will cut 1 mm thick mild steel at ~10 m/min and 10 mm thick steel at ~1 m/min.

The broad line graphs shown in this section give a range of cutting speeds at any particular thickness as an indication of the fact that the best quality cuts are rarely achieved at the highest speeds. Although cutting speeds are an important consideration when judging the feasibility of purchasing a machine they should be treated with a certain amount of caution. Most CO_2 laser users cut at between 80%–90% of the maximum rated cutting speed as the reliability of the process is extremely high under these conditions. It is also fairly unusual to run a laser continuously at its absolute maximum output power for the same reasons. Most CNC control machines have a certain dwell time before commencing an operation at a preset location and it is often necessary to program a dwell deliberately after the laser is initially fired in order to drill a hole through the material.

Taking into account these factors and such phenomena as CNC deceleration when turning corners, it is clear that the linear cutting speeds shown in this section are only indications of the process time for a real component. If, for example, a certain complex component involving a large number of small holes takes 100 s at a feed rate of 1 m/min, an increase in feed rate to 2 m/min may only decrease the cutting time to 75 s rather than the expected 50. The "floor to floor" time of the component will be even less affected as positioning and clamping times (if any) will not have changed.

A number of steel suppliers now produce grades of mild steel which are specifically aimed at the laser cutting market. Steels of this type have very low levels of impurities such as silicon which inhibit the cutting process. Although such steels have the same physical properties as the steels they replace, they can be cut 10–20% faster than ordinary mild steels. Also, the maximum thickness which can be cut by a particular laser will increase by a similar amount. These advantages are, of course, accompanied by an increase in the cost of the material.

The various designs of lasers and their cutting heads means that there is a range of cutting speeds available at any one power. The cutting speed depends on the laser mode, beam diameter, nozzle type and focusing lens focal length. It is these factors which combine to identify the energy input to the cutting zone and a full investigation of cutting speeds would therefore cover an enormous number of graphs showing the minor influences of each parameter. In practice, however, the majority of lasers sold for cutting purposes are axial flow machines with fairly similar modes using only a short range of lens focal lengths. The modes can be generally described as being circular in cross section with a general increase of intensity towards the centre, although in some cases the centre itself may be of slightly reduced intensity rather than a peak. Lens focal lengths are usually in the range of between 127 mm (5 in) and 64 mm (2.5 in) and nozzles are generally copper cones with a 1–2 mm diameter exit used at oxygen pressures of 1–4 bar.

As a result of this convergence of design throughout the industry, Fig. 2.3 can be taken as a good approximation of the cutting velocities possible from a laser which has been specifically sold as a cutting device. Lasers with modes which are not axially symmetric or which produce doughnut modes are not included in this group as they are not usually sold for cutting purposes.

2.1.3.2 *The Effects of Process Parameters on Cut Quality*

Laser cutting is a process within which a certain energy input to the cut zone is matched by energised material leaving that zone. The energy inputs in the case of mild steel are

1. The focused laser beam
2. The exothermic oxidation reaction
3. The mechanical force of the oxygen jet

These inputs and the time they are allowed to act in any one area are controlled by the following process parameters:

1. The laser power and spot size (power density)
2. The oxygen pressure and flow conditions
3. The cutting speed

These parameters should be balanced and set to levels at which they will ensure a steady removal of material from the cut zone. The consequences of an imbalance between the energy input and the material output from the cut zone are straightforward: if too much energy is put into the cut zone the material surrounding the intended cut line will become overheated and damaged; if too little energy is incoming the molten material in the cut zone will not be effectively removed.

Within certain boundaries it is possible to maintain the cutting equilibrium by treating the different energy inputs as equivalent. For example, if an increase in energy input to the cut zone is needed, it can be achieved by lowering the cutting speed, increasing the laser power or increasing the oxygen flow. Naturally this "equivalence" breaks down if it is taken beyond certain limits as only certain combinations of laser power, density, oxygen flow and process speed can achieve a cut. The factors governing these combinations can be more easily understood if the major input parameters are discussed separately.

The Laser Power Density. The laser power density is a function of the overall laser power and the size of the focused laser spot. These two variables can be discussed separately.

Laser power. This obviously has a direct influence on the energy input to the cut. It was mentioned earlier that in a typical application the laser would be set to cut at between 80% and 90% of the maximum cutting speed in order to make the process insensitive to process parameter fluctuations. If this is the case, a drop in laser power of ~10% will have a negligible effect on the cutting performance. Larger reductions in output power will result in incomplete penetration of the cut zone. If the "cutting" zone does not completely penetrate through the sheet of material, then the melt cannot flow out of the laser–material interaction zone in the usual way. As a result of the containment of the melt and the exothermic nature of the oxidation reaction, the workpiece becomes extensively overheated. This overheating can lead to sporadic penetration of the laser–oxygen jet combination.

On the other hand, a substantial increase in laser power can be easily accommodated by the cutting process. What tends to happen is that the majority of the surplus energy passes through the cut zone without interacting with the material. A quick check of the amount of energy being "wasted" in this way can be carried out by placing a sheet of wood or acrylic several centimetres below the cutting process and noting the amount of damage generated by the unused beam. If the laser power is too high for a particular job (e.g. 150% or more of the required power) then the overheating of the oxidised melt may give rise to burning at sharp corners.

Focus spot size and depth of focus. The diameter of the beam at the focused position can be established to a good approximation by drilling holes in thin section (~1 mm) acrylic sheet. It must be borne in mind, however, that the power density is proportional to $1/(\pi r_b^2)$ (r_b = beam radius) and so if a spot is twice the usual diameter its power density will be only one quarter of the smaller focused beam. For most cutting purposes the focused beam should be as small as possible in order to produce a cut by damaging the minimum amount of material. The size of the focused beam is proportional to the focal length of the lens used. Smaller focal length lenses therefore give a smaller more intense focal spot. Two drawbacks to using lenses of less than ~50 mm (~2 in) focal length are (1) the lens is obviously very near the cutting zone and can easily be damaged by spatter; (2) the depth of focus, or distance over which the beam can be assumed to be focused, becomes very small. For these reasons the most commonly used cutting lenses have a focal length of 63.5 mm (2.5 in) and 127 mm (5 in). Although laser manufacturers often quote the theoretical minimum focused spot diameter, the diameter of most actual focused spots lies between 0.1 mm and 0.4 mm. Much larger spots are produced by very high power machines (≥5 kW) and this makes them more suitable for welding than cutting. The beam is not focused down to a point but is converged down to a minimum "waist" as shown in Fig. 2.4. It is for this reason that we can talk about a depth of focus. The theoretical values for focused spot diameter and depth of focus often quoted by manufacturers are both smaller than those found in practice (see Sect. 10.6). The effective depth of focus for good quality high speed cutting varies from material to material. The range is largest for polymers (the nominal focal point can be positioned ~3% of the focal length of the lens above or below the sheet surface), is intermediate for steels (±~1%) and is smallest in the case of high reflectivity metals such as aluminium (±~0.5%). The +1% figure quoted here for steels is well within the limits of the process. Experimental work (Chap. 11, Ref. 35) has shown that using a 5 in (127 mm) focal length lens it is possible to cut with similar quality and speed over a focal range of 5 mm (±2% of the lens focal length). Generally it is wisest to find the nominal focal point (by methods explained in Sect. 5.3) and position this on the surface of the workpiece. In some cases, when cutting thicker section steels, improved results can be obtained by positioning the focal point two or three millimetres below the material surface.

These comments are all relevant to a machine which is working correctly

Fig. 2.4. The geometry of the beam in the focused region. The diameter of the beam does not reduce down to a point; it is converged to a minimum "waist" which can be almost constant over a few millimetres. The range of this small diameter beam is the "depth of focus". (As a rule of thumb the depth of focus can be assumed to be ~2% of the focal length of the lens.)

within the usual range of focused spot sizes. This spot size can be enlarged as a result of a number of faults in the system. Any such enlargement will mean a rapid reduction in beam intensity. In minor cases (e.g. radius increases by 10%) this beam enlargement may simply reduce the cutting speed and increase the kerf width. If the focused spot size is increased by a considerable amount, however, cutting will become seriously affected and may become impossible. Symptoms of such an effect would include widespread burning of the workpiece and a lack of cutting penetration. The usual system faults which could result in poor focusing characteristics are degradation in the mode quality, or scratches and spatter on the surface of the focusing lens or other optics.

Oxygen Flow. In Chap. 5, the section on the gas dynamics of nozzles demonstrates the complexity of the flow from a cutting nozzle. The oxygen flow through the cut zone adds chemical and mechanical energy to the cutting process. This dual role played by the oxygen jet is best understood by dividing it into a number of functions as follows:

1. Oxygen is supplied to the molten iron–iron oxide mixture on the cut front. The subsequent oxidation reaction is exothermic (heat generating) and this heat input helps to accelerate the cutting process. The iron–iron oxide mixture produced by this reaction has a low viscosity and a low cohesion to the solid material surrounding the workpiece. These factors are partially responsible for the efficient nature of melt ejection from the cut zone (see Chap. 10 for further details).
2. The mechanical force of the oxygen jet drives the melt out of the cut zone. This mechanism is obviously necessary to the cutting process but is in fact both an energy input and output as far as the cutting zone is concerned. The mechanical energy taken from the oxygen jet is, of course, an input, but this is far outweighed by the energy output associated with the removal of the highly energetic melt. It will be seen later that this energy input–output balance is responsible for the fact that too much or too little oxygen flow can both cause unwanted burning.
3. One final role of the oxygen stream which is often ignored is that of cooling by forced convection. This effect is negligible in the case of steels but can become noticeable when cutting non-ferrous metals or non-metals with air or inert gas. It is possible in these cases to reduce maximum cutting speeds by increasing the cutting gas jet flow.

The above points show that unlike the laser beam, which acts only as an energy input to the cutting zone, the oxygen stream is a major energy input (chemical and mechanical) and output (removal of hot material and convection). Excessive gas pressures result in corner burning as the chemical energy input to the cut zone outweighs the increased material removal capabilities of the high pressure gas jet. On the other hand, an insufficient gas jet pressure will also result in burning as material is no longer cleanly and quickly ejected from the cut zone. In this case the overheating of the cut zone is the result of the reduction in thermal output rather than an increase in input.

Process Speed. The process speed determines the length of time for which the laser–oxygen jet combination are at work in any particular area. The effect of changing the process speed is therefore to alter the energy input to the cutting zone at any particular point along the cut line. If the maximum cutting speed is exceeded then the cutting reaction cannot penetrate to the bottom of the sheet and very poor, incomplete cutting is the result. The effect is very similar to reducing the laser power as the energy input–output balance has been disturbed in the same way in both cases. Between the maximum cutting speed and the minimum there is a large operating window and cutting speeds can be reduced by up to 50% without deleterious effects. Once more the situation is analogous to changing the laser power, this time by increasing it. As in the case of cutting with too much power, the surplus energy passes straight through the cut zone without affecting the workpiece. At very low speeds, however, the workpiece can become overheated and burning can result.

2.1.4 Cut Initiation and Termination

2.1.4.1 Initiation

During cutting a burning front is established throughout the depth of the material. The energy, in the form of the focused laser beam and its coaxial oxygen jet, enters this cut zone from above and the oxidised, molten material leaves from the bottom. On the rare occasions when cutting is carried out from the edge of a steel sheet the cutting front is readily established by the interaction of the laser–oxygen combination with this edge. More usually, however, it is necessary to produce a hole in the sheet before cutting can proceed. The cutting machines which combine a laser with a punch press often employ a small radius circular tool to punch a hole for the laser to start from. The laser cut front is then generated by the same method as starting from the edge of a steel sheet. The more common "laser only" cutting machines employ the laser to produce the initial hole before cutting starts. The fundamental problem concerning the drilling of the initial hole is that until such a hole exists, the material can only be ejected from the top surface of the material back towards the laser.

Two basic methods exist:

1. Carefully controlled hole drilling using the laser in its pulsed mode and using air or oxygen as the drilling assist gas. This creates a small neat hole but needs careful CNC control and generally takes longer than the second method.
2. Simply firing the beam–oxygen combination at the selected spot using the same laser power and oxygen pressure which will be employed throughout the rest of the cut. This creates a hole almost instantaneously, but the hole size is greater than the kerf width and molten iron oxides in the form of sparks and small droplets are ejected in all directions.

These two techniques are best discussed separately.

Pulsed Drilling. During pulsed drilling of initial holes the high peak powers of the individual pulses serve to melt and vaporise a small amount of metal.

Large scale melting by conduction is frustrated by the short length of each pulse. Each pulse produces a small superheated melt which leaves the area as vapour and explosively ejected microscopic droplets. Air is often used as an assist gas to minimise the spreading of the hole by exothermic oxidation. In this case the term "assist gas" is rather misleading as the gas jet does not accelerate the drilling process, it merely acts to keep the nozzle and lens free from spatter. Once the sheet has been completely pierced in this way the assist gas is changed back to oxygen and cutting can proceed. For detailed work, including small diameter holes, this drilling process is superior to the blast method because the generation of the start up hole does not heat up the surrounding area and does not remove material which could later act as a heat sink to the cutting process. Obviously, a 1 mm diameter blast generated hole in the centre of a 2 mm diameter cut hole will, by conductive heating and material removal, severely upset the heat sinking capabilities of the material inside the 2 mm diameter circle and could result in unwanted material burning.

Blast Drilling. Exposing the material to a stationary focused beam with its coaxial oxygen jet will quickly generate a shallow weld which then exothermically oxidises. As the volume of melt grows, the oxygen jet will cause a depression in its centre and will then force the melt out of the area. This melt removal is rather explosive in nature and material is ejected radially away from the focused laser spot.

The size of the hole generated by this method is variable and can range from 100 μm diameter to 1 or 2 mm under identical processing conditions. The size of the hole is, to some extent, related to the thickness of the material being cut. An average blast hole has a diameter which is approximately equal to half the sheet thickness. The holes do not generally have parallel sides and are not necessarily circular. The size of these start up holes and the heat generated during their creation is partially responsible for the difficulty experienced when cutting holes in mild steel which have an aspect ratio of 1:1 or less (e.g. a 4 mm diameter hole in 4 mm thick sheet). Figure 2.5 demonstrates this very clearly. The two sets of photographs show the top and bottom faces of cut mild steel sheets of thicknesses of 6 mm and 1.6 mm. The holes cut in the 6 mm plate are 20, 10 and 6 mm diameter with a start up hole shown beneath these. In the case of the 1.6 mm sheet the holes are 6.0, 2.0 and 1.6 mm diameter with, once again, a start up hole at the bottom. In each case the hole with the 1:1 aspect ratio is clearly overheated and the cut edge has burned away from the intended line. Figure 2.6 shows the top and bottom views of typical circles removed during the cutting process. These are also from 6 mm sheet and represent the centres of circles 20, 10 and 6 mm diameter. Obviously, as the aspect ratio approaches 1:1 the start up hole takes up an appreciable volume of the circle. The local overheating and lack of heat sink results in wholesale melting on both sides of the cut line which spoils the shape of the required hole. In cases where holes have an aspect ratio of 1:1 or less, pulsed drilling of the initial hole is to be preferred to blast drilling. The size of the blast drilled hole can be reduced by employing air rather than oxygen as the assist gas but the drilling process will take more time (e.g. less than 1 s for oxygen, 2–3 s for air). One further method of minimising burning is to cool the blast hole before

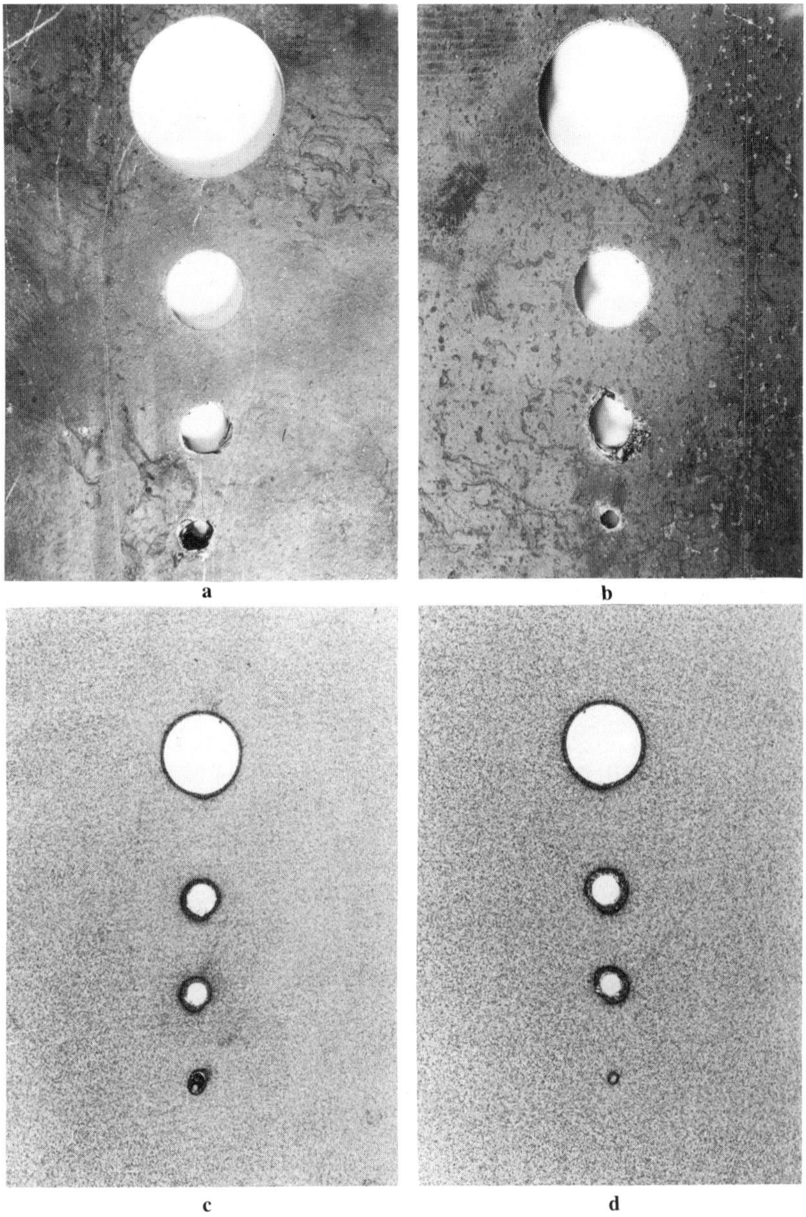

Fig. 2.5. The problem of burning around small diameter holes in mild steel: **a**,**b** Top and bottom views of 6 mm thick plate with holes of 20, 10 and 6 mm diameter (bottom hole is a typical piercing hole made to initiate cutting). **c**,**d** Top and bottom views of 2 mm thick plate with holes of 10, 3 and 2 mm diameter (with piercing hole below). Note how burning only becomes a problem as the hole diameter approaches the material thickness. Burning can be minimised by piercing all the start up holes in a component first and then going back to start the hole cutting after they have cooled (this cooling only takes a few seconds). Other methods include cutting with air rather than oxygen (at much lower cutting rates) and cutting with water coolant spray.

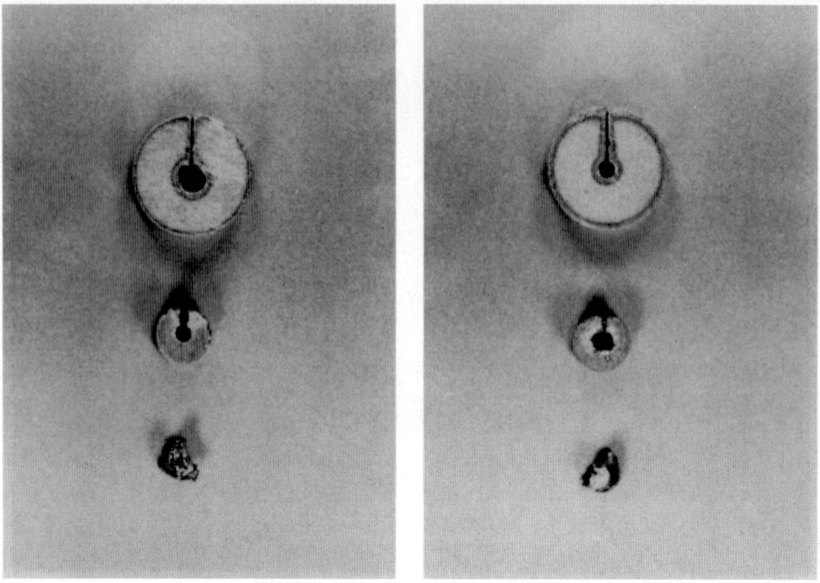

a b

Fig. 2.6. a Top and **b** bottom views of the circles cut out to produce the holes in 6 mm steel shown in Fig. 2.5a,b. It is clear that the smallest "circle" has been thoroughly overheated by the cutting process. This melting of the "circle" is accompanied by burning of the edges of the hole produced.

cutting is carried out. In some cases this cooling can be achieved by drilling a number of start up holes in a sheet and then returning to the first one to start cutting. The time interval between hole drilling and cutting gives the area time to cool down and thereby inhibits burning. Molten debris which sometimes accumulates around blast holes should be scraped away before cutting as this material can deflect the oxygen jet and disrupt cutting.

2.1.4.2 Cut Termination

Three features of the final stages of any cutting operation are worthy of note:

1. The cut is usually completed by meeting or crossing a previously cut line. If the time interval between the two cuts in the same area is not sufficiently large the workpiece may become locally overheated and burning may occur. (For notes on how to avoid burning see the following section.)
2. If the workpiece is not adequately supported the weight of the barely attached cutout can distort the line of the cut in the final stages.
3. In the case of all but the thinnest sections, the final millimetre or so of the cut will have a discontinuity of quality as shown in Fig. 2.7. This feature is a result of a change in material removal mechanism as the top leading edge of the cut zone finishes cutting before the trailing lower edge. The effect, which is due to the inclination of the cutting front, is

Fig. 2.7. The discontinuity of surface finish associated with the final stages of the completion of a cut (see also Fig. 2.8). In this case (10 mm thick mild steel) the disruption of cut quality is quite clear, although this effect may be negligible in thin sections (3 mm or less). (Cutting direction is from right to left.)

shown schematically in Fig. 2.8. It is clear that once the top part of the cutting front breaks clear of the workpiece then the cut zone becomes double sided to erode away the final "triangle" of connecting material. This effect is more pronounced in thicker section materials as the lag between the top and bottom parts of the cutting front is larger.

2.1.5 Techniques to Avoid Unwanted Material Burning

2.1.5.1 CNC Programming for Optimum Results

The following comments deal with simple techniques which have been developed to optimise the cutting process and to minimise potential problems, especially those associated with material overheating and burning.

Corner Overshooting. All shapes are made up of a series of angles and/or curves. Taking a typical 90° angle, two points are clear;

1. The movement system must slow down and momentarily stop to change direction if a true right angle is to be described.
2. The area included by the angle will become preferentially heated by the laser beam as the thermal diffusion away from the area becomes more difficult.

Most modern CNC machines are designed to minimise deceleration when turning corners, but it cannot be completely avoided. When mechanical machining operations are being carried out, this slowing down at corners does not generally affect the quality of the finished component. On the other hand, a dynamic thermal process such as laser cutting relies on a reasonably uniform cutting speed if the cutting mechanism is to remain in equilibrium.

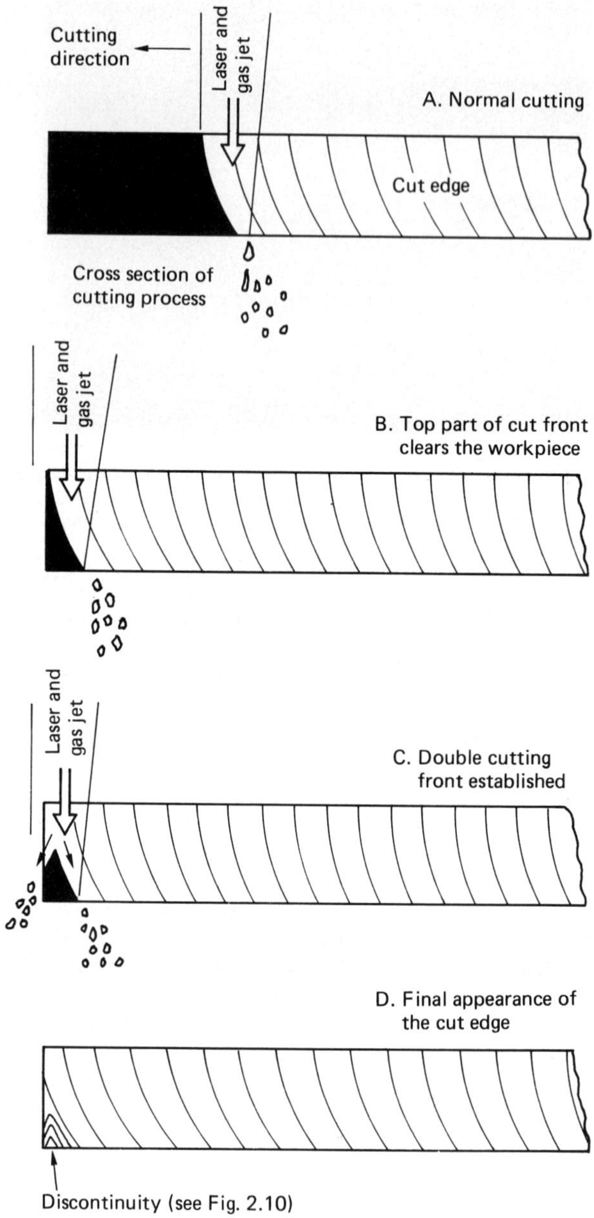

Fig. 2.8. A schematic of the generation of an end of cut discontinuity as shown in Fig. 2.7.

It is obvious therefore that when cutting mild steel with an oxygen jet in combination with the laser there will be a tendency for the area included by any angle to overheat. This overheating of "male" corners can lead to local burning away from the cut zone, an effect which becomes more pronounced as the included angle of the corner becomes smaller. To overcome this by

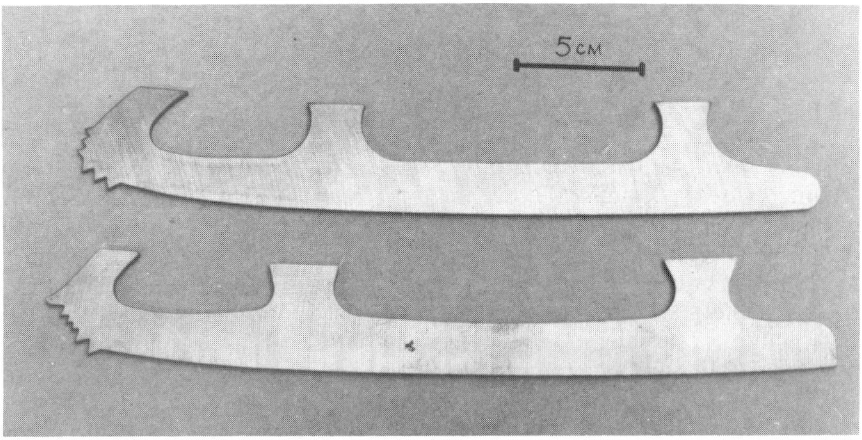

Fig. 2.9. Ice skate blades showing the toothed ends which are prone to burning if precautions are not taken.

programming methods is simple in many cases. Take, for example, the toothed end of an ice skate blade as shown in Fig. 2.9. If this shape is attempted directly at constant power the repeated deceleration of the beam together with the extra heating of the included angle zones act together to make the whole area around the tips of the points reach the ignition temperature in the oxygen jet. The resulting wholesale burning of the workpiece around the sharp details removes the tips from the teeth. One way to overcome this type of problem is to overshoot the sharp corners as demonstrated in Fig. 2.10a. Using this technique the teeth are generated by the intersection of two cuts. There is no deceleration around the point of the finished tooth and, although the heat sinking capabilities of the zone included by the two cuts are not as good as those on either side of a simple straight line cut, the material has had time to cool down between the first and second passage of the beam. An extension of this technique is demonstrated in Fig. 2.10b where a small radius circle has been used to overshoot the corner. This use of a circle has the same advantages of the rectilinear overshoots shown in Fig. 2.10a, but results in a smoother motion as abrupt changes in direction are avoided. Many CNC movement systems respond poorly to abrupt changes in direction which can give rise to a characteristic ripple after each corner. This ripple is generated by the machine com-

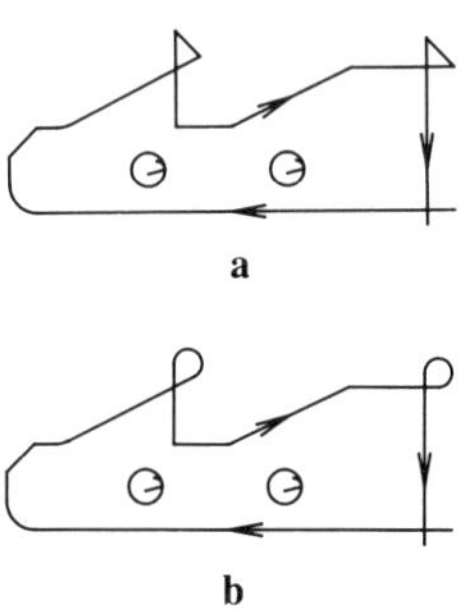

Fig. 2.10. Corner overshooting techniques to avoid burning. **a** Rectilinear overshooting; **b** circular overshooting.

pensating for its natural tendency to overshoot corners as a result of its momentum and the response time of the CNC controller. Use of a deliberate circular overshoot on corners can, therefore, not only eliminate burning, but can increase the geometrical accuracy of the cut shape.

Power Ramping. A complementary or alternative method to corner overshooting which helps to avoid burning on corners is power ramping. By this method the input power from the laser to the cutting zone is decreased to match the deceleration of the process at corners. In this way the equilibrium between the rate of energy input and the rate of material output from the cut zone is kept stable. As careful control of laser output power is required, the technique generally employs the pulsed mode of laser operation. Laser power can be easily ramped up or down by incrementally changing the pulse frequency or pulse length.

Some recent cutting machines employ preset power ramping subroutines to be used automatically on corners. On machines without this facility the correct power ramping conditions can be established after a few hours of trial and error experimentation.

More details of laser pulsing can be found later in Sect. 2.1.8.

Cutting Speed Ramping. Increasing the cutting speed has a similar effect to decreasing the laser power if the changes are kept within certain limits (typically 20%). It is therefore possible to compensate for the overall heating of the workpiece by increasing the cutting speed. An example of the possible application of this technique is during the final stages of many cuts where the laser is returning to a precut line.

Programming for Cutting Holes. In general, the start up hole of a laser cut should be contained in the waste material. (This is not necessarily the case if careful pulsed drilling of the start up hole is carried out; see Sect. 2.1.4.) If holes are to be cut out of a mild steel sheet the start up hole is therefore usually contained in the blank removed by cutting. The obvious route from the start up hole to the cutting path is a straight line, as shown in Fig. 2.11a but this can result in a definite mark where the cut starts and ends. Depending on the hole shape, the presence of stop–start marks on the cut edge can be minimised by using the type of cutting path outlined in Fig. 2.11b. These paths employ part circles to enter and leave the cutting line. This is, of course, in order to minimise the abrupt changes of direction which lead to localised burning (see previous section on corner overshooting). In the case of some small (e.g. 2 or 3 mm diameter) circles, this semicircular cut path entry might encourage burning of the component as the heat sink capabilities of the material contained within the circle become minimal towards the end of the cut. In this case, a straight line entry as in Fig. 2.11a may be the only answer unless power ramping or water cooled laser cutting (next section) are employed.

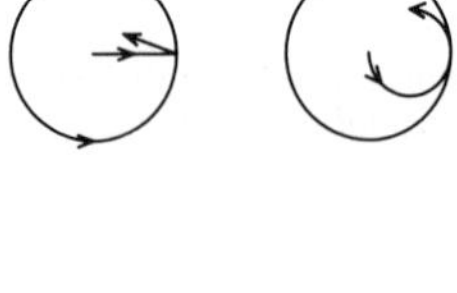

Fig. 2.11. Cutting paths for producing circular holes. **a** Straight entry and exit (can produce noticeable stop-start mark); **b** semicircular entry and exit (can lead to burning on small radius holes).

2.1.5.2 *Water Cooled Cutting*

One way to reduce unwanted burning of details is to cool the workpiece directly during cutting. This cooling is achieved by spraying the area around the cut zone with water or a water–soluble oil mixture. A fine water spray nozzle is used to deliver a circular "curtain" of coolant in a ring around the cutting zone. The pressurised oxygen cutting jet acts to keep the actual cut zone free of water but the recently cut edges are cooled by the spray. As the water does not actually interfere with the cut zone itself, the cutting speed is not affected.

The efficiency of the cooling process is enhanced by the fact that the water follows the cut slit as a result of capillary action. The removal of heat from the cut edges almost immediately after the passage of the cut zone prevents the workpiece becoming heated by conduction. As a result of this cooling, the thermodynamic equilibrium in the cut zone remains constant even if detailed work is being carried out. The flow of coolant can be as low as a few centilitres per minute for thin steel, although this may have to be larger for thicker sections. The rate of flow required is determined by the onset of widespread boiling of the coolant. If a very low mass flow rate of water is used, a substantial amount of water vapour can be generated. This vapour is unpleasant for the operator and can cause damage to the worktable and external laser optics. As the flow is increased this boiling process becomes inhibited and vapour generation is minimised. The coolant used in the system can be either water or a mixture of water and soluble oil. The oil solution has the advantage that corrosion of the workpiece is avoided during subsequent storage of the damp cut components. When water alone is used it is necessary to dry the components to avoid rusting in storage. Most oil–water systems employ recirculation of the coolant which involves collecting the coolant, filtering it and then pumping it back into the cutting zone spray.

Although the primary use of water cooling is the elimination of burning of details when cutting mild steel, the method can also be employed to reduce heat affected zones in carbon steels and to minimise oxidation discoloration around cuts in stainless steel. Water cooled cutting has also been used to good effect when cutting non-ferrous metals such as aluminium and titanium. In the latter case the water spray not only minimises workpiece overheating but also helps to trap and wash away the large volumes of oxide dust generated during laser–oxygen cutting of titanium alloys.

2.1.5.3 *Cutting with Air*

If air or diluted oxygen is used as the cutting gas, burning of the workpiece is eliminated as a result of the suppression of the exothermic oxidation reaction and the fact that the air stream acts as a coolant by forced convection. As a result of the suppression of the oxidation reaction the kerf width is reduced, typically from ~0.3 mm to ~0.15 mm. There are a few drawbacks to the use of air, of which a minor one is the presence of dross on the lower edge of the cut. This dross is the result of the incomplete oxidation of the melt produced during cutting. The unoxidised iron has a high surface tension and remains firmly attached to the substrate. The more major arguments against the use of air as a cutting gas is the large reductions necessary in both the cutting speed and the maximum thickness of the workpiece. Cutting speeds can be reduced to 10% or 20% of the oxygen cut value. The maximum thickness of steel which can be cut can be reduced by 50% or more, compared with oxygen cutting. Heat affected zones tend to be broader than those associated with oxygen cutting as a result of the prolonged laser–material interaction time at any point along the cut.

In spite of these limitations, laser–air cutting of steels is a valid method for producing fine detailed work. Perhaps the most common use of an air jet in the context of steel cutting is not for the cutting itself but the generation of small, neat start up holes (see earlier section on cut initiation).

If a computer controlled gas switching system is available it is possible to combine oxygen and air cutting on one component to improve productivity. It should be borne in mind that the best quality, highest speed, air assisted cuts will be achieved at substantially higher pressures than those used for oxygen. Typical air pressures will range from 3 to 10 bar (50–150 psi) depending on material thickness (air pressure should be increased with material thickness). Some laser cutting heads are not designed to withstand pressures in excess of 5 bar (~70 psi) and some redesigning may be necessary, particularly where lens diameters are greater than 25 mm (1 in). This may, however, be simply a matter of improving "0 ring" seals and buying thicker section lenses (without changing the focal length).

The best air supply for cutting is from compressed air bottles. Air from compressors tends to have the following drawbacks:

1. Maximum pressures are low.
2. The pressure fluctuates by ~10%–20%.
3. The air usually contains oil and always contains water.

Water and oil can contaminate the lens in the cutting head which will then start to absorb the laser beam. This absorption not only detracts from the cutting power, it can lead to permanent damage and lens fracture. It is also worth noting that as the lens warms up by beam absorption it swells and changes its focal length. This phenomenon can interrupt the cutting process as the focal point of the beam rises above the material surface.

If compressor air is used for cutting or for keeping external laser optics clean it must be filtered several times. The final filter should be a carbon based odour remover and the system should also include a method of drying the air.

2.1.6 The Effect of Sheet Surface Finish on the Cutting Process

Mild steel is available with a variety of surface finishes. At thicknesses below approximately 4 mm (0.16 in) the steel is reduced down to thickness by cold rolling which gives it a clean bright surface. At thicker sections the steel is rolled in its hot condition and the finished material has a dark grey or black appearance. This black surface coating is made up of oxides picked up during the rolling process and can have an effect on cutting characteristics.

Mild steel is often zinc coated or painted to prevent corrosion. The nature of laser cutting means that it can be carried out on semifinished products which may have such an anti-corrosion surface. Depending on the type and thickness of such coatings it may be necessary to adjust the process parameters used for cutting. The following discussion will comment upon the most common surface characteristics of mild steel in the following order:

1. Bright (cold reduced or cold rolled)
2. Black (hot rolled)
3. Pickled and oiled
4. Zinc coated (galvanised or hot dipped)
5. Painted
6. Shot blasted
7. Rusty (corroded)

2.1.6.1 Bright (Cold Reduced or Cold Rolled)

This heading has been included in the list merely as a datum point to which the subsequent coatings can be compared. Cold reduced, bright mild steel has no surface coating of any sort. The material may, however, be covered in a thin layer of oil to prevent corrosion during storage and transportation. The vaporisation of this oil around the cut zone can interfere with the cutting process and is also an irritant to the operator. For these reasons it is best either to degrease the sheet, or at least to wipe excess oil off with a cloth. This second course of action is often preferred as it is obviously quicker and cheaper than wiping with a degreasing agent, and also a certain amount of oil is retained on the cut components to retard corrosion during storage.

2.1.6.2 Black (Hot Rolled)

It is common to reduce the section of steel from a billet to sheet by passing it between rollers while the material is in its yellow- to red-hot state. At these elevated temperatures the steel is very ductile and rolling can be carried out at a high rate. Throughout the process the hot steel sheet is exposed to the air and to the water used as a lubricant and coolant. It is not surprising therefore that the finished sheet is covered in a substantial layer (~0.1 mm (0.004 in) thick) of oxides which have been produced during production and simultaneously embedded into the material surface. This surface coating is traditionally known as mill-scale.

The mechanism of laser cutting for mild steels relies on the exothermic oxidation of iron as a major energy source in combination with the laser. In the case of hot rolled steel the surface is oxidised before laser cutting is carried out and therefore does not contribute its energy of oxide formation to the cutting process. The thickness of the surface layer of oxide varies considerably from batch to batch. In the case of a thick oxide layer on a thin section sheet, the presence of the pre-oxidised material can dilute the

amount of moisture. It is possible that the combination of the iron oxides and its associated water will inhibit the cutting process. The vapour given off by the boiling water can interfere with the oxidation reaction by diluting the oxygen stream. The resultant cuts are of rather poor quality, showing evidence of sporadic burning and dross adhesion. For this reason rust should generally be cleaned off steel before laser cutting is attempted.

2.1.7 The Importance of Axial Symmetry of the Energy Input to the Cutting Zone

It is clear that during the cutting of mild steel a delicately balanced dynamic equilibrium is established rather than a "steady" state where a continuous input of energy is matched by a continuous flow of material out of the cut zone. The striation generation reactions take place in a circular manner around the centre line of the movement of the laser over the workpiece surface (see Sect. 2.1.2). It is therefore very important that the energy input to the area is axially symmetric (i.e. identical in cross section in all directions). The axial symmetry of the energy input can be affected in any of four ways:

1. The symmetry of the laser mode can be imperfect due to poor tuning or damaged optics.
2. Any linear polarisation of the beam can be considered as asymmetry as the beam will cut better in certain directions than others.
3. The symmetry of the oxygen jet can be disturbed by nozzle damage or contamination.
4. If the symmetry of the oxygen jet and the laser mode are individually very good, the symmetry of their combination can be upset if the two are not coaxial, due to incorrect centring of the nozzle with the beam.

These types of asymmetry are the reason behind the sensitivity of the process to mode quality and nozzle condition mentioned in Sect. 2.1.1. Any lack of symmetry will result in inferior cutting in certain directions. Symptoms of this inferior cutting will take the form of increased cut edge roughness, adherent dross on the lower lip of the cut edge, material burning at corners and reduced cutting speeds. These types of energy input asymmetry will be dealt with separately in the following sections.

2.1.7.1 Symmetry of the Laser Mode

An asymmetric mode will cut with varying degrees of effectiveness depending on the direction of cutting. The non-symmetric nature of the energy distribution of the beam will, in effect, mean that the focused beam will have some areas which are "hotter" than others. Cutting in different directions at a set speed will result in varying cutting conditions, depending upon where the hot spots within the mode are with respect to the direction of cutting. The delicately balanced thermodynamic reactions in the cut zone have their reaction kinetics determined by the incident local energy density.

If, for example, the beam has its greatest energy density just behind its centre when cutting in a certain direction, it may cut with excellent quality. On the other hand, when cutting in the opposite direction the maximum energy density will be slightly in advance of the beam centre, cutting dynamics will change, and it may cut badly, producing a rough edge. Cutting in a direction perpendicular to this axis, the "hottest" part of the beam will be off to one side which will give rise to preferential ignition of that side in the oxygen jet. This situation will give rise to widely different cut edge qualities between the two sides of the cut.

The following experiment was carried out to show these effects clearly:

1. A laser with a good Gaussian output was used to cut 4 mm thick mild steel in eight different directions at a power of 500 W. The directions of cut were labelled N, NE, etc. for convenience.
2. The laser was detuned to give an asymmetric mode and the eight cuts were repeated. The loss of power due to the detuning was compensated for by turning up the electrical supply to the laser.
3. All the cutting parameters were kept identical for both sets of cuts with the exception of the mode quality.

The modes in each case were recorded by mode burns and by use of a laser beam analyser (see Chap. 11, Refs 63 and 64) which gives an oscilloscope trace of two cross sections of the mode at 90° to each other. The cuts produced were examined photographically and by profilometry.

Figure 2.12 shows the energy distributions of the two modes as revealed by acrylic mode prints and the laser beam analyser. The degradation of the symmetry of the beam is quite clear. The effect of this corruption of the

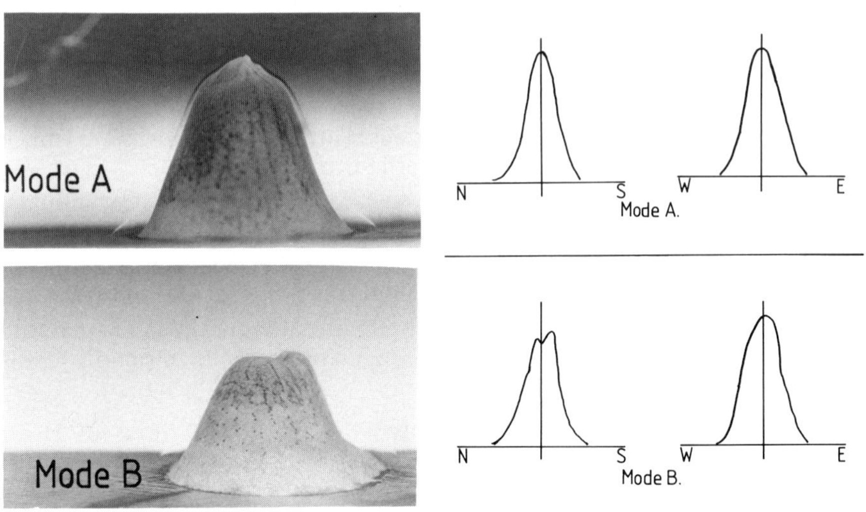

Fig. 2.12. The energy distribution of modes A and B defined by **a** mode prints in acrylic; **b** laser beam analyser (see Sect. 5.5 for notes on both beam analysis methods).

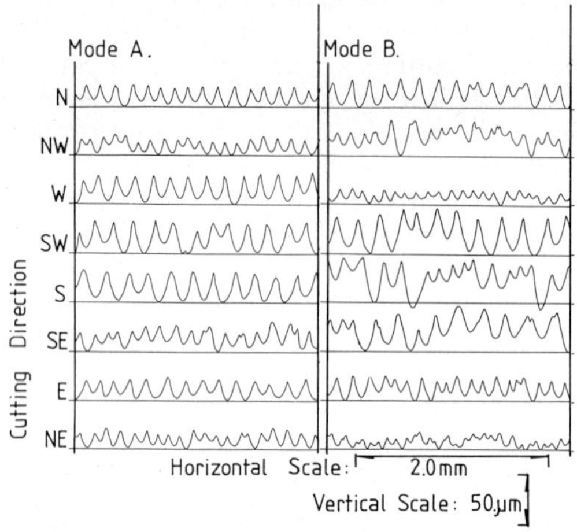

Fig. 2.13. Magnified profile traces of the cut edges produced in eight different directions by the two modes.

beam on the cutting process is demonstrated clearly in Fig. 2.13 which shows the magnified profiles of the cut edges.

The cutting results of the modes are discussed separately.

Mode A. The profiles relating to Mode A cutting reflect the axial symmetry and near Gaussian nature of the energy distribution in the beam. The laser material interaction is, in all directions, strongly cyclic in nature, producing cut edges of similar, low, roughness with no adherent dross in any case. Kerf widths in all cutting directions were 180 μm (±20 m).

Mode B. Comparison of the samples and profiles for Mode B with those of Mode A revealed the following points:

1. The differences between the various cutting directions were much greater as far as cut quality is concerned.
2. The surface roughness and general surface quality was inferior in many cases, particularly in cutting directions N, S and SE.
3. Adherent dross was plainly evident on the lower edges of most of the cuts. The most prominent cases being directions SW, SE, and NE.
4. In the case of the rougher cuts, the periodic nature of the laser–material interaction (as indicated by the regularity of the striations) was disturbed.

Differences in cut quality between this set of results and those for Mode A can be attributed solely to the change in mode as great care was taken to replicate all the other process parameters (laser power 500 W, cutting speed 1.0 m/min, oxygen pressure 1.5 bar).

The delicate thermodynamic and fluid dynamic balance which is required for the cyclic laser–material interaction which generates the cut has obviously been disturbed in the case of Mode B. It is clear that a different balance is achieved in each of the eight cutting directions as a result of the reduction in axial symmetry in the beam. Kerf widths in this case were 5% to 10% greater than the 180 μm for Mode A. This increase in kerf width is attributable to the "flattening" of the mode peak which would result in broader focused spot diameter. The general reduction in energy density caused by the lateral spread of the mode and its inferior focusing characteristics has resulted in a drop in the effectiveness of the cutting process, which in turn has resulted in adherent dross on the lower edge of the cut. Sporadic deep burning of the cut edge, particularly when cutting in the north–south direction, can be attributed to the increased laser–material interaction time as the larger focused spot moves through the cut zone. This increase in laser heating time at lower energy densities can lead to localised overheating of the material and therefore excessive burning in the oxygen jet.

A follow up experiment which involved further detuning the laser (but keeping the power constant at 500 W) resulted in very poor quality cutting in a number of directions. It was also found that cutting was not possible in some directions unless the cutting speed was reduced by 20%.

From these points it is clear that axial symmetry of the laser mode is extremely important when cutting mild steel with a CO_2 laser. Many other materials are less "mode sensitive", but in any case mode symmetry assists the uniformity of cutting performance in all directions.

2.1.7.2 Polarisation of the Laser Beam

The subject of polarisation and its relevance to laser cutting is a rather complex one and will be dealt with here in a simplified conversational manner. Readers who require more information are advised to read the work of Olsen on the subject (Ref. 25, Chap. 11).

Every photon or "light particle" is made up of an electrical and a magnetic vector at right angles to each other which can be represented as in Fig. 2.14a. The light emitted from most CO_2 lasers is polarised as a result of the geometric design of the laser cavity. "Polarised" means that all the photons have their electrical and magnetic vectors aligned parallel to each other, as shown in Fig. 2.14b. This alignment of the vectors means that the light beam as a whole has highly directional properties. The beam is better absorbed in certain cutting directions than others and this effect gives rise to a symptom which is specific to polarised beams. This symptom is the sideways inclination of the cut front from the line of propagation of the beam unless the cutting direction is parallel to the polarisation axis, i.e. if the beam is pointing vertically the cut can be inclined by a substantial angle as shown in Fig. 2.15. The clearest demonstration of this problem is to cut a circle in mild steel. The top of the cut disc will be geometrically accurate but the bottom will be elongated and elliptical.

This polarisation of the beam has to be removed before the beam will cut equally well in all directions. The depolarisation is carried out by reflection off one or two specially coated mirrors between the laser and the work piece. The coating on the mirrors has the effect of "twisting" the photons so that the vectors which are at 90° to the direction of propagation follow a circular pattern, as shown in Fig. 2.16, rather than all pointing in the same

high volume liquid oxygen tank.

2.1.7.4 *Misalignment of the Nozzle*

Nozzle misalignment is one of the most common reasons for poor cutting quality as the process is extremely susceptible to any discrepancy in the alignment of the cutting gas jet with the laser beam. The gas flow from the nozzle generates a pressure gradient on the material surface which is, of course, coaxial with the nozzle itself. The focused laser beam establishes the position of the cutting zone and this will naturally lie directly under the central core of the gas jet only if the nozzle and beam are coaxial.

When a cut is established the gas can move away from the high pressure core of the gas jet in two ways;

1. Vertically down the cut front
2. In a lateral direction across the sheet metal surface away from the maximum pressure zone at the centre of the gas jet

Figure 2.17a shows the equilibrium set up if the gas jet and the laser beam are coaxial. If, however, the gas jet is not coaxial with the laser as in Fig. 2.17b, the movement of gas away from the centre of the gas jet produces an overall flow across the top of the cut zone. This lateral flow can entrain material from the cut zone to generate a shower of sparks travelling across the sheet metal top surface. The direction the sparks take identifies the axis of the misalignment and shows the direction in which the nozzle (or the lens) must be moved to eliminate the problem. The generation of a secondary shower of sparks is a clear symptom of misalignment and will be accompanied by poor cut quality (dross adhesion, burning etc.). Photographs of the secondary spark shower and the associated cut edge quality are given in Chap. 6 (Figs 6.1 and 6.2).

Misalignment of the nozzle can also be identified by a noticeable reduction in the major spark shower leaving the bottom of the cut when cutting in certain directions. This is generally caused by the fact that the central core of the gas jet is slightly (e.g. ~0.2 mm) in advance of the cut zone. If the gas jet centre is slightly behind the cut front cutting is not greatly affected and so the material removal process is only disrupted when cutting in one direction. This direction should be identified and the nozzle adjusted to alleviate the problem (see Chap. 6).

Summary. From these notes it is clear that the axial symmetry of the energy input to the cut zone is of prime importance when cutting mild steel. Most other materials are not as sensitive to minor disturbances in symmetry and will cut well under conditions which would badly affect mild steel. For this reason the cutting of mild steel can be used as a test for any laser cutting machine. A good quality machine should be able to cut mild steel at speeds similar to those in Fig. 2.3. The maximum thickness which can be cut by a particular machine should be the number equal to ~1% of the laser output power expressed in millimetres (e.g. 500 W; 5 mm, 800 W; 8 mm). This rule of thumb is approximate and only works between 200 and 1500 W.

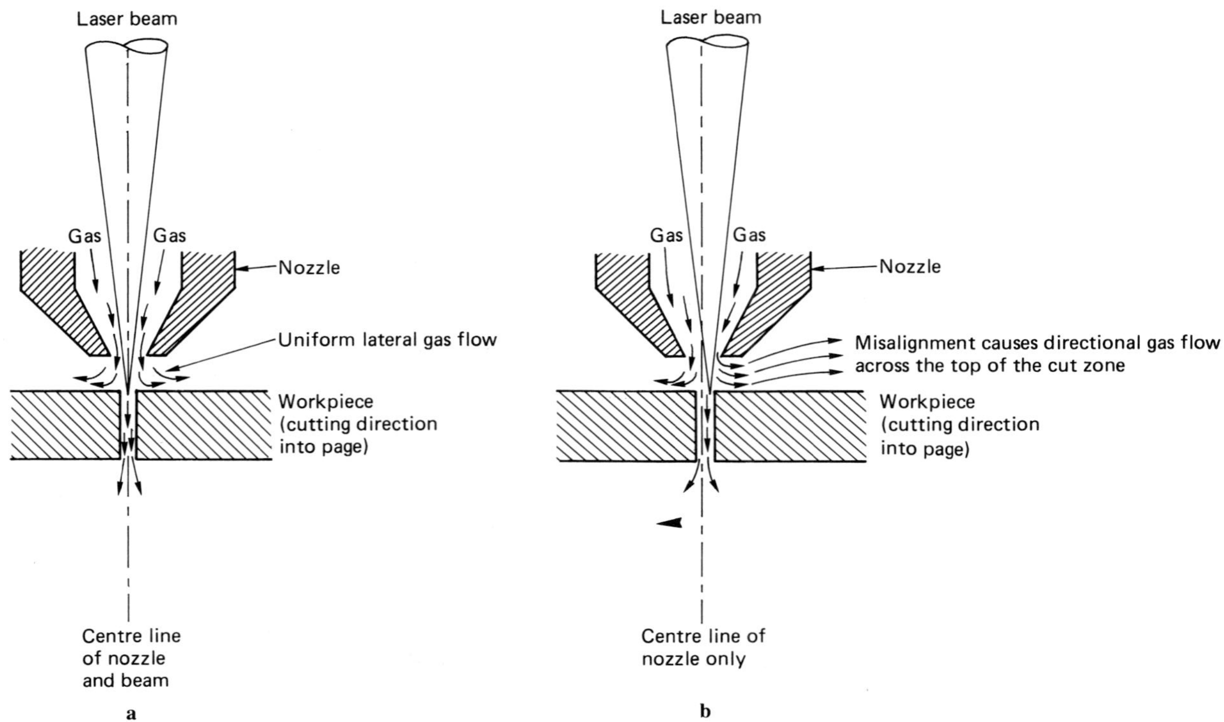

Fig. 2.17. **a** The equilibrium set up when the gas jet and laser beam are coaxial. **b** Nozzle–beam misalignment leads to an overall gas flow across the top of the cut zone which can lead to unwanted cut edge burning. (See also Fig. 5.8.)

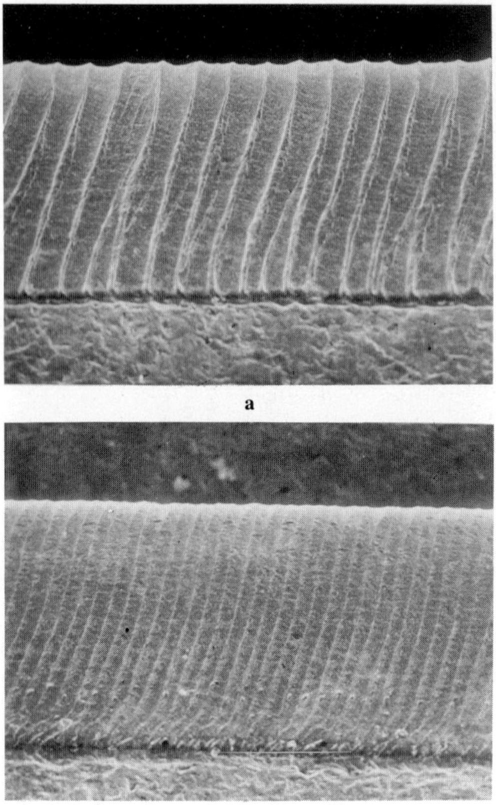

Fig. 2.18. A comparison of **a** continuous wave laser cutting and **b** pulsed laser cutting (pulse frequency 500 Hz, on:off ratio 4:1). All other cutting conditions were kept constant for the two samples, i.e. material thickness 1.25 mm (0.05 in), average laser power 300 W, oxygen pressure 2.0 bar, cutting speed 1.8 m/min (70 in/min). The roughness of the pulsed sample (Ra) was only 25% of the continuous wave sample.

2.1.8 Pulsed Laser Cutting of Mild Steels

2.1.8.1 Introduction

The earlier section on the mechanism of cutting for mild steels showed that the surface roughness of the cut is largely determined by the nature of the striations generated along the cut edge. When cutting with a continuous wave (c.w.) laser output, the pitch and amplitude of these striations is controlled by a number of factors most of which are specific to the material itself (e.g. thickness, alloy content etc.). It is possible, however, to change the nature of the striations on the cut edge by pulsing the laser beam. The basic idea is to produce striations by repeatedly drilling overlapping holes through the material. One pulse of laser energy is used to drill one hole and thus generates one striation. In this way the control over the pitch and amplitude of the striations moves from the material characteristics to the laser machine. Using the correct pulsing parameters it is possible to produce cut edge striations which are very close together, and this in turn means that the cut surface quality improves because the roughness diminishes. A clear example of the improvements possible in practice is shown in Fig. 2.18. The two cut edges compared here were cut under identical conditions of cutting speed, average laser output power, oxygen jet pressure, etc. but in the case of the edge shown as Fig. 2.18b, the laser output was pulsed at a frequency of 500 Hz. The technique works best in cases where a clear striation pattern covers the cut edge from top to bottom, i.e. for thin sections. Thicker section steel has a clear striation pattern at the top of the cut edge which becomes obscured by the flow of liquid metal towards the bottom of the cut. In these cases pulsing can improve quality but, as individual pulses become incapable of drilling through the thickness of the material, the results will not be as obviously rewarding as those shown in Fig. 2.18.

The technique of pulsing which will be discussed in this section is a method of improving the surface finish of a cut, but a more common industrial use of pulsing is to reduce burning on corners or detailed work. Many CNC laser cutting machines now encorporate a facility whereby the laser switches from continuous wave to pulsed mode when going round small radii or corners. In this case the pulsed mode is used because it is very easy to change the power input to the workpiece to compensate for any deceleration on the part of the laser–material movement system. The use of a stream of individual pulses rather than a continuous input of c.w. radiation minimises the build up of heat in the workpiece and therefore prevents burning. Cutting machines which have this facility are available in a range of levels of sophistication, but quite often it is sufficient to employ a single pulse setting which is engaged for all corners. The control over the overall power input to the workpiece when using the pulsed mode is far more flexible than simply altering the output power in the c.w. mode, because the pulse length, frequency and shape can all be individually altered.

This method of pulsing to eliminate corner burning involves a reduction in the overall average output power of the laser in accordance with the requirements of the reduced cutting speeds at corners. The technique discussed next differs from this approach as the idea is to maintain the same average laser output power and the same cutting speed as c.w. cutting, whilst at the same time improving the cut edge quality.

The Technique. Depending on the attitude of the people involved and the equipment available to them, there are two methods of investigating laser pulsing. The first method involves trial and error within certain given boundaries and can often generate impressive results within two or three minutes although the best possible parameter settings could be missed. The second method is much more scientific and meticulous and is probably more applicable to a research and development environment. Both methods will be described here separately.

2.1.8.2 The Trial and Error Method

1. Cut a thin section (~2 mm, 0.08 in) sheet of mild steel using the laser in

its continuous wave mode. Make a note of:

The laser power
The cutting speed
The material thickness
The oxygen flow settings

2. Switch the laser over to the pulsed mode (some instruction from the manufacturers will be needed even if this takes the form of simply reading the laser handbook). Set the pulse frequency to 500 Hz and the pulse length to 1 ms (see notes 1 and 2 below).

3. Using the current control which is generally used to change the laser power in c.w. mode, increase the power until it is equal to the earlier noted c.w. power (see note 3 below). If this power level is not achievable at 500 Hz and 1 ms increase the pulse length, but do not exceed 1.8 ms. (Obviously 500 Hz at 2 ms means that the laser is executing one long overlapping pulse which is equivalent to c.w. output.)

4. Having achieved the earlier noted average output power repeat the cut at the same speed and oxygen conditions as earlier. Visually compare the two cut edges.

5. Increase the pulse rate to 550 Hz and decrease the laser power to the earlier noted c.w. value by decreasing the pulse length. At this frequency do not exceed a pulse length of 1.6 ms. (1.82 ms = c.w.). Once again cut and compare.

6. Repeat 5 for pulse rates of:

600 Hz: max pulse length ~1.5 ms (1.67 ms = c.w.)
650 Hz: max pulse length ~1.4 ms (1.54 ms = c.w.)
700 Hz: max pulse length ~1.3 ms (1.43 ms = c.w.)
750 Hz: max pulse length ~1.2 ms (1.33 ms = c.w.)

Above a frequency of ~750 Hz there is a tendency for the steel to return to c.w. cutting behaviour because it is incapable of responding fast enough to the individual pulses. This phenomenon might be experienced at higher or lower frequencies depending on the pulse shape of the laser output.

7. For material with a thickness of greater than 2 or 3 mm it is possible to use lower frequencies than 500 Hz as the lower cutting speed means that even low frequency pulses will be close together on the finished cut edge. Maximum pulse lengths are as follows:

450 Hz: maximum pulse length ~1.8 ms (2.2 ms = c.w.)
400 Hz: maximum pulse length ~2.0 ms (2.5 ms = c.w.)
350 Hz: maximum pulse length ~2.3 ms (2.8 ms = c.w.)
300 Hz: maximum pulse length ~2.8 ms (3.3 ms = c.w.)
250 Hz: maximum pulse length ~3.5 ms (4.0 ms = c.w.)
200 Hz: maximum pulse length ~4.5 ms (5.0 ms = c.w.)
150 Hz: maximum pulse length ~6.2 ms (6.7 ms = c.w.)
100 Hz: maximum pulse length ~9.0 ms (10.0 ms = c.w.)

NB: At low frequencies the average power may drop and speeds must be reduced accordingly.

Notes. 1. *Pulse frequency*: It may be necessary to calibrate the pulsing system. This is best carried out by a service engineer from the laser manufacturer but an approximate estimation of the pulse frequency can be obtained by passing material rapidly under the focused beam. The oxygen supply should be replaced by an air jet to minimise damage to the material as the aim is to leave a trail of individual "dots" across the material surface. The pulse frequency can then be estimated by counting the dots per unit distance and, from the materials speed of travel, calculating how many dots (pulses) per second were being generated, e.g. If the dot separation is 1 mm and the material speed was 500 mm/s then the pulse rate is 500 Hz.

$$\frac{\text{Material speed}}{\text{Dot separation}} = \text{Pulse frequency}$$

If the *XY* movement system has a maximum velocity which is too low to allow the dots to be clearly separated, a spinning disc placed under the beam can be used to achieve higher speeds. The linear velocity (material speed) of such an arrangement can be easily calculated from the following formula:

$$V = \frac{\pi r f}{30} \quad \text{or} \quad V = 0.105\, rf \quad \text{(mm/s or in/s)}$$

where

V = the linear velocity (mm/s or in/s)
r = radius of circle of dots generated (mm or in)
f = rotation speed in revolutions per minute (r.p.m.)

(Obviously the units used for the measurement of r will determine the units in which V is expressed.)

One quick guideline to the pulsing frequency is simply to listen to the laser; 500 Hz sounds like a fairly high pitched hum, 1000 Hz sounds like a whine etc.

2. *Pulse length*: It is important to separate the pulses emitted from the laser in order to generate a truly pulsed beam. With many laser systems it is easy to set the pulse generator incorrectly so that the pulses overlap each other. For example, pulses of 2 ms duration cannot be generated at 1000 Hz as this would involve 2 s worth of "pulse on" time per second. The pulse width in milliseconds multiplied by the frequency in kilohertz must always be less than one. When the width multiplied by the frequency is equal to one, the arcs in the laser are starting one pulse discharge immediately at the end of the previous one which is effectively continuous operation. For this reason the pulsing experiment should always begin at a pulse on/off ratio of 1:1, e.g. 1 ms at 500 Hz (= 0.5 s of "pulse on" time per second). If the average power required cannot be achieved with this amount of "off time" the pulse width should be gradually increased but should remain below a maximum where ~90% of the available time is "on time". Beyond this level of pulse width the material will be unable to distinguish between individual pulses and will revert to c.w. cutting behaviour.

3. *Output power*: Some lasers do not have a user operable means of changing the current control. If this is the case the output power will be determined by the pulse length for a given frequency. Therefore, set the

pulse length to the maximum which will allow individual pulses (~1.8 ms for 500 Hz). Cutting speeds may have to be reduced to compensate for the lower power available from pulsing, but the cut quality will probably be improved.

2.1.8.3 The Scientific Method

1. If possible, introduce yourself to the subject by studying Refs 32 and 33 (Chap. 11).

2. To carry out work of the highest quality the following equipment will be required:

storage oscilloscope and camera
Laser Beam Analyser (Refs 64 and 65, Chap. 11)
low powered optical microscope
surface profilometer (e.g. Talysurf)

3. The process is basically that outlined in the trial and error method except that the observations of the pulse characteristics and the cut qualities are more detailed and systematic. After the first (c.w.) cut is carried out it is useful to count the number of striations per millimetre by using the microscope. From the wavelength of the striations and the cutting velocity the frequency of striation generation can be ascertained, e.g. 10 striations per millimetre at a cutting speed of 30 mm/s gives a striation generation frequency of 300 Hz.

The aim of the technique is to maximise the striation generation frequency in order to pack the striations as close as possible together and thus achieve the smoothest cut edge. As a starting point attempt to produce striations at twice the naturally occurring frequency for c.w. cutting. In the case above this would, of course, mean setting the laser pulse frequency at 600 Hz. The pulse frequency, width and shape can be monitored using the laser beam analyser and the storage oscilloscope. By systematically varying these three parameters but keeping the overall average power constant it is possible to fine-tune the process to produce the best possible results.

2.2 Alloy Steels

2.2.1 Introduction

Alloying elements are added to iron in order to give a range of materials with a variety of properties. The most commonly required properties are strength, hardness, toughness, corrosion resistance and specific magnetic characteristics. Iron alloys have been developed to satisfy the requirements of each of these properties individually and in various combinations. For example, high levels of hardness and strength can be achieved by adding up to 1.0% of carbon to iron and then executing the correct thermal cycle. Alloy steels range in alloy content from 1%–2% (as in the case of carbon steels) up to stainless steels with alloy contents of over 30%. The various alloying elements and their concentration have different effects on the laser cutting process and therefore need to be discussed under separate headings. Rather than discuss each element individually, it is convenient to divide the following discussion into sections dealing with the most common groups of iron alloys:

Stainless steels
Low alloy steels
Silicon iron
Free cutting steels
Carbon steels

2.2.2 Stainless Steels

2.2.2.1 Grades of Stainless Steel and Their Applications

As the name suggests, stainless steels were developed to resist corrosion. The major corrosion resistant alloying addition is chromium which, unfortunately, inhibits the laser cutting process by the same mechanism by which it prevents corrosion. The frustration of oxidation in either case, is achieved by the chromium initially reacting with oxygen to produce a surface layer of chromium oxide (Cr_2O_3) which forms an impermeable seal over the underlying material. Stainless steels all contain more than 10% chromium with or without other alloying additions. There are four main groups of stainless steels; Austenitic, Ferritic, Duplex and Martensitic; these terms refer to their internal crystallographic structure.

These four types of steel are differentiated by their alloy content and the uses to which they are put as follows.

Austenitic Stainless Steels. Steels in this group are undoubtedly the most important members of the stainless steel family and are the most commonly used and laser cut. They possess a high resistance to corrosion, good weldability, toughness at subzero temperatures and excellent ductility.

Austenitic stainless steels used nowadays differ only marginally from the first of the type which was developed in 1923 and contained 18% chromium and 8% nickel. For this reason austenitic steels are often referred to as "18/8" stainless steel. Although these materials cannot be strengthened or hardened by thermal treatments their high corrosion resistance has ensured their use in the following areas; brewery, dairy and chemical plant, storage tanks, sinks, saucepans, control panels, display and signwork, filters and a multitude of other applications.

Austenitic stainless steels can be rapidly identified by the ease with which they can be scratched and by their lack of response to a magnet. Naturally these tests also hold true for other materials such as aluminium and should not be considered foolproof.

Ferritic Stainless Steels. The ferritic stainless steels contain 17% or 20% chromium but without the addition of nickel or any other major alloying ingredient. As they have a good ductility, these materials are used principally as cold rolled sheet and strip for subsequent press or roll forming. The lower chromium content (17%) ferritic sheets are used for such applications as motor car trim and washing machine parts. The higher chromium

(20%) grades are more resistant to corrosion at elevated temperatures (up to 900 °C) and are commonly used for oil burner nozzles and cooker grill frets, etc. Compared with austenitic stainless steels, these ferritic grades have a poor weldability and are unsuitable for use at low temperatures. Their corrosion resistance is also inferior to austenitic steels, but because of the lower alloy content, they are not as expensive. Ferritic and martensitic stainless steels cannot be distinguished from bright mild steel by the use of a magnet as they all respond in a similar manner.

Duplex Stainless Steels. Duplex stainless steels have an internal chemistry which falls between those of austenitic and ferritic. They have intermediate properties to match this intermediate chemistry and are used in applications where a strength superior to that of austenitic is to be combined with a corrosion resistance superior to that of ferritic.

Martensitic Stainless Steels. Martensitic stainless steels differ from the other two main types in that they can be hardened and strengthened by suitable heat treatment. Hardnesses and strengths can be achieved which are over twice those of austenitic and ferritic materials. Chromium contents are lower than for the other grades (13%–16%) and other alloying additions are minimal if anything (e.g. 2.5% N with 16% Cr). Carbon is present in higher quantities than in the other stainless steels, for example 0.1%–0.8% rather than 0.03%–0.1%. Although these materials have advantages as far as strength and hardness are concerned the weldability, formability and corrosion resistance is poor compared with the ferritic and austenitic grades. Applications of these materials include ball and roller bearings, cutlery, surgical instruments, valves, shafts and general engineering fittings.

2.2.2.2 *The Mechanism of Cutting*

In the case of all stainless steels the laser cutting mechanism is basically the same. A melt is generated throughout the depth of the workpiece and this is partially ejected from the cut zone by the incident oxygen jet. The situation in the cut zone is shown schematically in Fig. 2.19. The oxygen reacts exothermically with the iron, chromium and any other alloying elements, but the process is not quite as effective as it is for mild steel. During mild steel cutting the oxidation reaction generates heat and a low viscosity melt of iron oxide which has a low compatibility with the solid iron at the edges of the cut zone. The oxide is thus easily removed as it has no strong attachment to the rest of the workpiece. In the case of stainless steels, however, the chromium in the workpiece affects the cutting process in a number of ways:

1. Under the action of the oxygen jet the chromium forms a layer of chromium rich oxide on the surface of the melt. Oxygen cannot pass through this Cr_2O_3 layer and further oxidation of the melt is frustrated. The chromium is preferentially brought to the surface of the melt as a result of its high affinity for oxygen even when compared to that of highly reactive iron. The presence of the Cr_2O_3 does not, however, completely prevent further oxidation as it is a liquid surface on a turbulent melt. The movement of the melt under the action of the

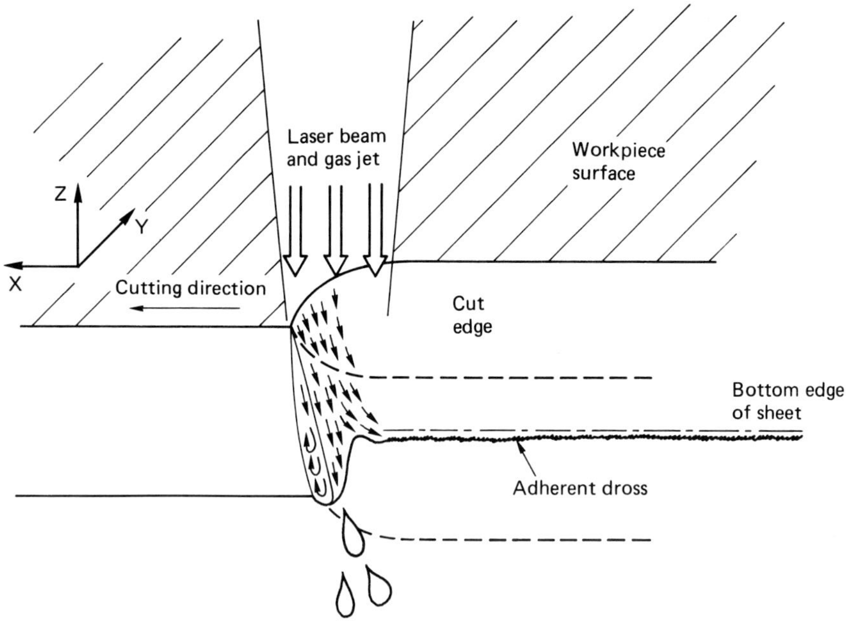

Fig. 2.19. A schematic of the cut zone when cutting stainless steel. The high surface tension of the melt and its adhesion to the workpiece results in dross adhering to the lower edge of the cut.

oxygen jet continuously ruptures this protective outer layer and allows the passage of oxygen. Another major factor which prevents the chromium oxide from producing a good "seal" over the underlying melt is the fact that it is diluted with iron oxides. One thermodynamic benefit which results from the presence of chromium is the increased energy of oxidation generated, compared with that of iron (see Chap. 10).

2. Even though the continuous fracture of the Cr_2O_3 "skin" on the melt surface allows some further oxidation to take place, there remains a substantial amount of unoxidised metal in the cut zone. This melt is expelled from the cut zone by the pressurised gas jet but a residue is left behind attached to the lower edge of the cut. The bulk of this residue is unoxidised and is therefore strongly attached to the parent metal and cannot be entirely removed by the action of the oxygen jet.
3. The surface tension of the Cr_2O_3 covered melt is greater than that of the more extensively oxidised melt produced during mild steel cutting. This is another factor which inhibits melt removal and results in residual melt or "dross" on the lower edge of the cut.

The cutting mechanism is, therefore, a combination of the oxidation cutting experienced by mild steel and melt shearing. Melt shearing is the mechanism by which titanium is cut in a jet of argon or thermoplastic polymers are cut in air (see Chaps 3 and 4). During melt shearing cutting a liquid pool is established throughout the depth of the material. This pool is

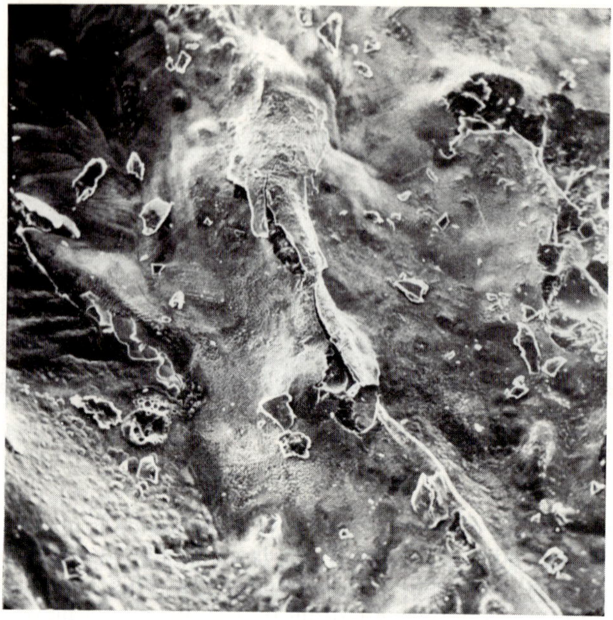

Fig. 2.22. An electron micrograph showing a magnified view of a laser–oxygen cut stainless steel edge. The thin surface layer (1–5 μm thick) of oxide can be seen to be extensively cracked. (Scale: ×300.)

Table 2.2. Chemical composition of a typical austenitic "18/8" stainless steel

	Fe	Cr	Ni	Mn	Mo	Nb	Si	C	S	P
Weight %	70.00	17.66	9.30	1.57	0.24	0.7	0.39	0.04	0.01	0.025
Atomic %	69.35	18.77	8.76	1.58	0.14	0.42	0.77	0.18	0.017	0.04

preferential oxidation the liquid below the outer oxide layer is, naturally, depleted of its chromium content. This point was investigated by Auger spectroscopy which established that the chromium content of the resolidified molten zone could be reduced to ~17% (atomic). This drop from ~19% to ~17% constitutes a decrease of ~10% of the chromium content. This decrease affects the corrosion resistance of the cut edge which will become coated in iron oxide (rust) if exposed to moisture.

It is clear from this discussion that the cut edge produced in stainless steels consists of the three different layers as shown in Fig. 2.23. These can be described as follows:

1. An outer layer of chromium oxide and iron oxide which is hard, brittle and extensively microcracked.
2. A rapidly resolidified melt which is chromium depleted when compared to the substrate.
3. The substrate material itself.

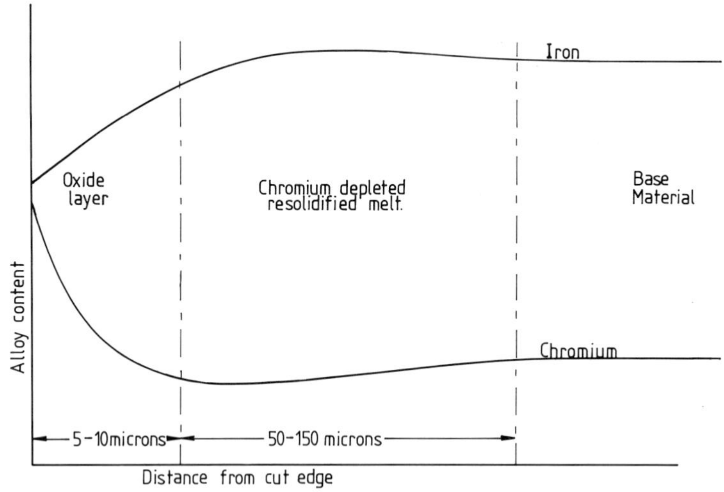

Fig. 2.23. A schematic of the changes in chemistry across the cross section of a stainless steel cut edge.

In the case of non-austenitic steels there will be a fourth zone between 2 and 3 which is made up of heat affected zone which has been hardened by the thermal cycle associated with cutting.

Laser–oxygen stainless steel cutting is a well established and important part of the laser cutting industry which obviously satisfies the requirements of a great many users, usually without a post-cutting cleaning operation. In certain applications, however, the condition of the "as cut" edge can pose problems to a subsequent fabrication operation or to the service performance of the component:

1. Welding of laser–oxygen components can be difficult if a high quality weld is required. The oxides on the cut surface tend to become trapped in the weld metal and remain as inclusions which reduce the tensile strength and fatigue life of the weld. (NB: If this or any of the following points is a serious problem they can be alleviated by using high pressure inert gas techniques, although product on costs will be much higher (see next section).)
2. In corrosive environments the reduction in the chromium content of the resolidified metal under the cracked oxide layer can lead to corrosion. This is a surface effect only and, although it can lead to unsightly staining it does not indicate a degradation of the material as a whole.
3. If a laser cut edge is in sliding contact with another component it will initially wear that component more quickly than a mechanically cut edge would. This is, of course, due to the hard nature of the Fe_2O_3–Cr_2O_3 surface. As this brittle surface is broken off the softer stainless steel beneath will be revealed and wear conditions will return to normal. It must be borne in mind, however, that if the oxidised surface is worn off it acts as a source of very hard grit which could damage other moving components.

4. The most common problem associated with a laser cut stainless steel edge is the physical presence of the dross itself. The dross obviously stands proud of the cut edge and this can lead to geometrical problems on construction of the final component (e.g. difficulty in bolting one sheet tight against another).

In many cases the problems reviewed above can be alleviated by cleaning the cut edge by grinding, wire brushing or rumbling. One, in process, technique is to eliminate the dross from the cut component by use of an auxiliary gas jet (see Sect. 3.2.2.2). Dross can be minimised by using very high pressure (≥6 bar) oxygen jets but this can give rise to difficulties associated with the strength of the lens which forms a part of the pressurised chamber supplying the nozzle.

2.2.2.5 High Pressure Inert Gas Cutting

Very high quality cuts can be produced by using a high pressure inert gas jet rather than the more usual oxygen jet. The resulting cut edge is dross free and unoxidised. This absence of oxidation means that the cut edges will have the same corrosion resistance as the bulk material. Also the cut edges may be welded without any post cutting preparation. Commercially this type of cutting is generally known as clean cutting.

Nitrogen is the gas most commonly employed although it is not completely inert. The reasons for the choice of nitrogen are its low chemical activity and its cheapness compared with truly inert gases such as argon. Figure 2.24 shows a typical good quality edge produced by high pressure inert gas cutting. Gas pressures employed are often in excess of 15 bar (~220 psi) and so the laser cutting head must be of a suitable design to cope

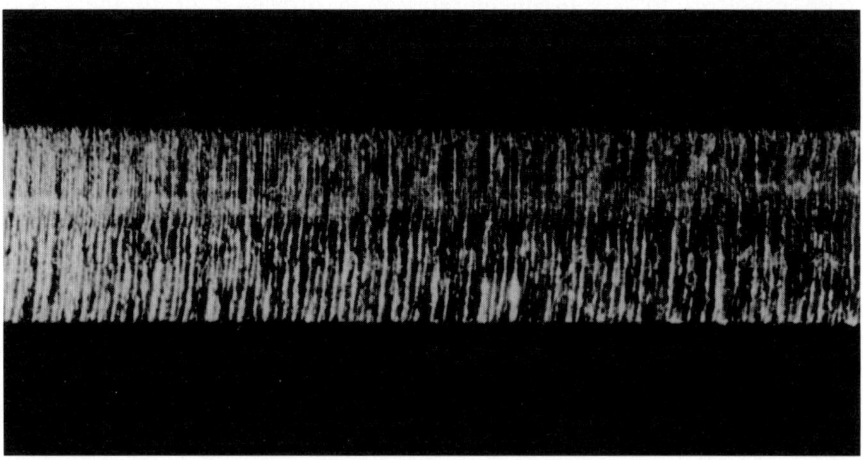

Fig. 2.24. The "as cut" stainless steel edge produced by high pressure nitrogen cutting. Although the quality is superior to laser–oxygen cutting, production costs are up to three times higher. Cutting parameters in this case were: 5 mm thick stainless steel cut at 1.1 m/min (43 in/min) with a laser power of 1.4 kW and a nitrogen pressure of 14 bar. The beam was focused in the middle of the section of the sheet (i.e. 2.5 mm below the top surface). Photograph courtesy of Dirk Petring (see Refs 26–28, Chap. 11).

with this. One particular weak point in the pressure system is the lens.

High pressure lenses should be several millimetres thick right up to the edge and be of a minimum diameter. This minimum diameter is obviously limited by the diameter of the incident beam. Typically, the minimum diameter of the lens will be approximately 150% of the diameter of the beam.

During normal laser cutting a minor amount of dirt or spatter on the lens can be accommodated. This is not the case in high pressure cutting and any level of surface contamination can lead to the lens bursting. This lens disintegration can cause damage to the components in the optical chain from the laser to the workpiece and, in any case, a thorough clean up will be required. If laser powers of less than 1.5kW are used, cutting speeds will be lower than those associated with oxygen cutting except at the thinnest sections (eg 1mm). This naturally has the effect of increasing the cost of the cut product. With the recent advent of lasers with higher powers and high beam quality, this problem is disappearing. 3kW machines can cut, for example, 2mm thick stainless steel with nitrogen at the same speed or even faster than by the oxygen assisted process. If oxygen is used when cutting thin section material it is common practice to reduce the output power of 3kW machines to improve the cut quality. If nitrogen is used as the cutting gas the power can be left at its maximum and higher cutting speeds can be achieved. As the stainless steel thickness is increased the oxygen assisted process can also use the highest powers and becomes faster than nitrogen cutting. At the moment (1998) stainless steel cutting is divided between the two types of assist gas. In the future it is likely that nitrogen cutting with higher power lasers will dominate this part of the market. For further reading on the technique the work of Petring et al. (see Refs 27 and 28, Chap. 11) is recommended. It is worth mentioning at this point that greater thicknesses of stainless steel can be cut by high pressure inert gas cutting than are possible by oxygen cutting, although speeds are very low.

2.2.3 Low Alloy Steels

The demands of various engineering applications have led to the development of a large number of alloys with physical properties which are superior to those of mild steel. The previous section reviewed stainless steels which were developed primarily for corrosion resistance. Other physical properties such as high strength and shock resistance are made possible by the addition of very small alloying additions. These "low alloy" materials include structural steels, spring steels, tool steels etc. The total alloying addition to the iron is usually less than 5%. For example, a typical chisel steel will consist of 0.4% carbon and 2% chromium with 97.6% iron.

Many structural steels are produced with alloying additions within the ranges given in Table 2.3, with specific levels of the various alloying components giving specific characteristics. This compositional range is given merely as an example to show that only minor alterations of the chemistry of the steel can give a wide range of properties.

Table 2.3. Typical alloying addition ranges for most structural steels

Element	C	Si	Mn	Cr	Ni	Mo
Wt%	0.25–0.45	0.1–0.35	0.35–0.9	0–1.5	0–3.0	0–0.7

In general these low levels of alloying elements have only a minor effect on the cutting characteristics of the steel. The metallic elements can be considered to be contaminants which interfere with the iron oxidation reaction with varying degrees of effectiveness. In general, cutting speeds may have to be reduced, compared with the speeds for mild steel. Dross may also present a problem but deposits will not be as heavy as those found on stainless steels. In some cases the material will cut at much the same speed as mild steel but with a superior cut edge. This improvement of the cut quality is a result of the suppression of the mechanism by which striations are produced on the cut edge of mild steel. Low alloy steels which cut in this manner often have a loose, oxidised dross deposit on the lower edge of the cut. This material can be easily removed by wiping, as it is not firmly attached to the workpiece.

2.2.4 Silicon Iron

The addition of a small amount (4%) of silicon to iron improves its performance in electrical applications such as transformers and motors. The silicon helps to reduce wasteful eddy current losses by raising the electrical resistivity to about four times that of pure iron. Another effect of silicon as an alloying element is to improve the resistance of steels to high temperature oxidation. It is not surprising, therefore, that silicon has a deleterious effect on the cutting process (compared with mild steel) which results in lower speed cuts with a dross covered lower edge. The reasons for this drop in cut quality and cutting speed are not clear, but it is apparent that the surface tension of the melt in the cut zone is increased and that the oxidation of the iron is somewhat frustrated. One possible reason for this is the fact that silicon dioxide and the iron–silicon oxide which are formed during the cutting process have a much lower density than the iron oxides. It is probable, therefore, that these silicon oxides form a surface layer on the melt which will affect both the surface tension and the oxidation kinetics. The relative densities of the oxides are shown in Table 2.4.

Table 2.4. The relative densities of silicon–iron oxides (Ref. 78, Chap. 11)

Oxide	Melting point (°C)	Density at ambient temperatures (kg/m^3)
FeO	1370	5700
Fe_3O_4	1565	5240
Fe_2O_3	1594	5180
SiO_2	1700	2500
$FeSiO_3$	1146	3500
Fe_2SiO_4	1503	4340

As many of the applications of silicon iron involve the close stacking of a number of thin plates or laminations, the presence of dross is extremely undesirable. The problem can be alleviated by using the pulsed cutting technique discussed in Sect. 2.1.8. Pulsed cutting reduces any segregation of oxides as melting is not a continuous event. One other method of dealing with the problem of dross is the use of auxiliary gas jets discussed in Sect. 3.2.2.2.

2.2.5 Free Cutting Steels

Free cutting steels have small amounts of manganese and sulphur added to them to improve their machinability. Although these elements have a beneficial effect on the response of the material to mechanical cutting tools, they can interfere with the laser cutting process. Typical compositions range from 0.1% C, 0.3% Mn and 0.08% S to 0.5% C, 1.6% Mn and 0.33% S. At the low end of the range, cutting is not substantially different from mild steel cutting, but as the alloy content increases the cutting process can be disrupted. The reason for this disruption is the generation of SO_2 and CO_2 as the alloying elements react with the oxygen stream. Although the oxidation reactions of sulphur and carbon are exothermic and therefore contribute energy to the cut zone, they generate gases which then dilute the oxygen reacting with the cut front. It was mentioned earlier (Sect. 2.1.1) that the oxidation process in the cut zone is very susceptible to small levels of contamination of the oxygen jet. The result of this "in process" contamination by the evolution of SO_2 and CO_2 is a reduction in cutting speeds and cut edge quality. In the case of the highest carbon and sulphur levels the cut edge may be covered in adherent dross in the same way as stainless steel (see Sect. 2.2.2).

2.2.6 Carbon Steels

Mild steel can be rendered harder and stronger by the addition of up to approximately 1% of carbon in conjunction with the correct heat treatment. Examples of the application of various carbon content steels are given in Table 2.5.

Table 2.5. Applications of carbon steels

Carbon content (%)	Typical applications
0.05–0.15 (dead mild)	Automobile body panels, tin plate, stampings, materials for drawing or pressing
0.1–0.2 (mild)	Boiler plate, joists, screws, forgings, general engineering applications
0.2–0.3 (low carbon)	Machines, structural work, gears, shafts and forgings
0.3–0.5 (medium carbon)	Axles, connecting rods, crane hooks, forgings, crankshafts, die blocks, turbine rotors
0.5–0.9 (high carbon)	Locomotive tyres and rails, wire ropes, band saws, hammers, drop hammer dies, punches
0.9–1.2 (tool steels)	Axes, drills, taps, screw dies, files, razors, boring tools

As the carbon content of the material increases the following effects can be observed:

1. Cutting speeds are reduced as compared with those for mild steel, e.g. 0.9% carbon steel ("ground flat stock" or "gauge plate") will cut at approximately 75% of the speed of mild steel.
2. The maximum thickness which can be cut is reduced, e.g. for 0.9% carbon steel the maximum thickness will be approximately 75% of the mild steel value.
3. The cut edge quality will change. Apart from the heat affected zone considerations which will be discussed shortly, the cut edge will have two distinctive features:
   (a) The striation pattern on the edge will be on a finer scale and may look superior to mild steel.
   (b) The lower edge of the cut may be covered in a dark powdery dross which is easily removed by wiping one piece against another.

All these effects are the result of the combustion of the carbon in the oxygen jet. This combination generates heat but also produces the gas CO_2 which acts to dilute the oxygen at the cut front and thereby inhibit the oxidation reaction (see Sect. 2.1.1). This inhibition may be so severe in the case of the highest carbon content steels that cutting is prevented altogether. Higher oxygen pressures can be employed to improve the situation by washing out the CO_2 but this can give rise to localised burning.

Even if a high carbon steel component is cut to a high visual inspection standard it must be borne in mind that there are some important metallurgical side effects associated with the cutting process. The severe heating–cooling cycle experienced by the workpiece in the areas on either side of the cut line will give rise to Heat Affected Zones (HAZs) similar to those produced by welding or heat treatment.

Heat Affected Zones. It was mentioned earlier that a certain carbon content steel gives the required strength and hardness only if it is subjected to the correct thermal cycle. A typical cycle for a high carbon steel would be:

1. Heat the steel to above ~800 °C and then cool rapidly by quenching in water or oil to give the maximum hardness.
2. Heat to a temperature in the region of ~250 °C for a substantial time (e.g. 1 h) to reduce the internal stress of the material. This "tempering" operation reduces the hardness somewhat but has tremendously beneficial effects on the toughness of the material.

Carbon steels are not usually supplied in this "hardened and tempered" condition because the material would obviously be difficult to work. It is more usual therefore to receive sheet material in its annealed or "soft" form. Traditional machining or bending operations can then be carried out as the steel responds in a similar way to mild steel. Once the component has been shaped it can then be hardened and tempered to give the required physical properties.

This state of affairs is not designed to incorporate a thermal cutting process such as laser cutting which will automatically harden the region close to the cut edge. The effectiveness of laser cutting depends on the overall chemistry of the material rather than its crystal structure (or hardness condition). It is therefore just as easy to cut the material in its hardened state as it is to cut annealed sheet. This can be a great advantage of the process which will undoubtedly be made more use of in the future. At the moment, however, it is more usual to cut high carbon steel sheet in its "as delivered" annealed condition. This may or may not cause problems to subsequent manufacturing steps. For the majority of applications the laser profiled sheet constitutes the completed component which then only needs to be heat treated. In this case the thermal effects of the laser cutting will disappear during the traditional heat treatment thermal cycle. In other cases, where the cut sheet must be machined or bent, problems can arise as a result of the hardened HAZ near the cut edge.

The thermal cycle of cutting produces an extremely hard edge as the heated metal is "quenched" by the surrounding cold material. This quenching is far more effective than that achieved by using oil or water coolants. The heat treated edge is hard and strong but also brittle. This means that this zone will damage any mechanical cutting edge and also resist any bending operation. If bending is successful it may be accompanied by cracking of the cut edge HAZ which may lead to failure of the component in future service. To avoid such problems the laser cut workpiece must be annealed before bending or machining. It may be more sensible, however, to laser cut the profile after all the other shaping operations are completed.

Certain high carbon steel alloys can crack spontaneously if exposed to a severe thermal cycle. This phenomenon may generate microcracks along

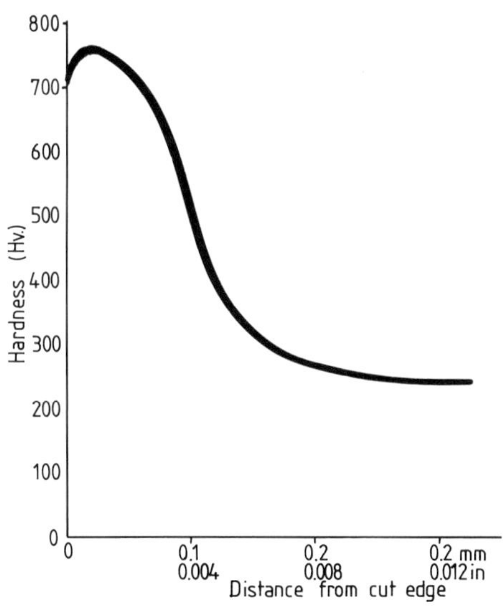

Fig. 2.25. A typical hardness profile of the edge of a high carbon steel cut component (2 mm thick material, hardness measured along centreline, 1 mm from top surface). The material reaches a maximum hardness value close to the edge. The extent of the hardened zone increased with material thickness. A "rule of thumb" would be to assume that the extent of the hardened zone is approximately one quarter of the thickness of the material.

laser cut edges which would reduce the fatigue life of the finished component. In most applications this would not be a problem, but critical components should be microscopically inspected and the alloy changed if necessary.

In certain limited applications the hard edge produced by laser cutting can be used directly, thereby avoiding the expensive step of heat treatment. The cut edge could, for example, provide sufficient hardness for a sliding wear operation with the soft, ductile bulk of the material providing the toughness. A number of different edge surface qualities and hardness profiles can be provided by cutting in any of the following ways:

1. *Cutting with oxygen*. In this case the cut edge quality will be similar to that of a mild steel component but the thermal cycle will generate a hardened edge. A typical hardness profile is given in Fig. 2.25. The depth of the hardened zone will change from a minimum near the top surface of the cut to a maximum at the bottom. This is because the bottom of the cut zone experiences a more prolonged thermal cycle than the top. The extent of the hardened zone will also increase if the thickness of the material is increased because cutting speeds will be reduced. As a rule of thumb the depth of the HAZ will be of the same order of magnitude as the thermal discolouration on the top and bottom surfaces of the cut component along the cut edge. This will generally be between one quarter and a half of the thickness of the material.
2. *Cutting with oxygen using a water cooling spray* (see Sect. 2.1.5.2). This will produce a cut edge which looks the same as in 1 but has a narrower hardened zone. This narrower zone may be only half as deep as the non water jet version but it generally still needs to be removed or heat treated.
3. *Cutting with air*. This produces a smoother edged cut with a lower edge covered in adherent dross (which can be removed by grinding). Cutting speeds are reduced by a factor of four or so, and this means that the thermal cycle at any point is prolonged. The longer thermal cycle generates a deeper, more gradual heat affected zone which may extend to a depth equal to the material thickness. The overall component will be stronger than the ones cut by oxygen and this method can be used to produce ready hardened components.

Chapter 3

Cutting Non-ferrous Metals

3.1 Introduction

Non-ferrous metals are, as the name implies, those metals and alloys of which iron is not a primary constituent, although it may be present in small quantities. The term covers a multitude of alloys with a variety of histories from bronze to the modern super alloys. This chapter will deal only with the four groups of alloy which are commonly cut by CO_2 laser, which are those based upon titanium, aluminium, nickel and copper.

These materials can all be cut with great technical and commercial success by CO_2 lasers, but the cutting mechanism is not as efficient as that already described for steels. Cutting speeds, which will be given in the following sections, can be up to an order of magnitude lower for non-ferrous metals. This reduction in cutting efficiency is a result of a general increase in reflectivity and thermal conductivity linked with a less efficient oxidation reaction than that employed when cutting iron based alloys.

It was seen in the previous chapter that mild steel is very effectively cut as a result of the exothermic reaction of iron which generates a low viscosity, easy-to-remove oxide, whilst at the same time providing the cutting zone with a great deal of heat. Of the four groups of alloys listed above, only titanium reacts in much the same way as iron and can be cut with the same level of efficiency. This is, however, misleading as the oxidised edge produced by this method is generally unacceptable on a titanium alloy finished product. High speed laser–oxygen cutting of titanium alloys can only be used as a rough cutting process and for a few restricted applications where the reduction in fatigue life associated with edge oxidation is unimportant.

Some of the basic differences between iron and the non ferrous metals under discussion are given in Table 3.1. The absorptivity value given describes the proportion of the incident light which is absorbed at room temperature. (In this table absorptivity as a physical property is preferred to reflectivity because the magnitude of the difference between reflectivities of 98% and 99% is not as clear as that between the associated absorptivities of 2% and 1%.) The absorptivity of any metal rises strongly with temperature and with the levels of surface oxidation often generated during cutting, so the values given in Table 3.1 are for comparison only.

The aim of all laser cutting is to generate a localised hot region where material is melted or otherwise thermally degraded to produce a cut. Any material property which inhibits the energy input to the cut zone or

Table 3.1. Selected properties of various metals (unpolished)

Metal	Absorptivity[ab] %	Thermal conductivity[a] W/m/K	Melting point K	°C	Heat generated during oxidation[a] kJ/mol	
Gold	0.5–1.0	296	1340	1067	–	–
Silver	0.5–1.0	419	1230	957	–	–
Copper	1.0–2.0	385	1356	1083	160	(CuO)
Aluminium	1.0–2.0	201	932	659	1670	(Al_2O_3)
Lead	~4.5	35	600	327	276	(PbO_2)
Iron	~5.0	80	1810	1537	822	(Fe_2O_3)
Nickel	~6.0	59	1726	1453	244	(NiO)
Titanium	~8.5	23	1950	1677	912	(TiO_2)
Tin	~16.0	65	505	232	581	(SnO_2)

[a] Values for pure metals at ambient temperatures (for comparison only).
[b] Absorptivity (%) = 100 − Reflectivity (%).

increases the energy output from that zone will reduce the effectiveness of laser cutting. For this reason the speeds of cutting possible are related to the material properties, particularly reflectivity (which inhibits laser input to the cut zone) and conductivity (which effectively cools the hot cut zone). Materials with high reflectivity (low absorptivity) and high thermal conductivity, such as copper and aluminium, cut slower than materials with lower values for these material properties. Also the maximum thickness which can be cut will be lower for a particular laser power.

A further influence on the efficiency of the cutting process is the energy available from the oxidation process. It is clear from Table 3.1 that copper and nickel evolve only a small amount of energy during oxidation, compared with iron or titanium. Aluminium, on the other hand, is highly exothermic and from this consideration alone, high cutting speeds would be expected. Unfortunately the achievement of these high cutting speeds is prevented as a result of a fundamental difference between the oxides generated by iron and those produced by aluminium. Unlike iron oxide(s), aluminium oxide (Al_2O_3) generated by the combined action of the focused laser and oxygen jet forms an impervious seal on the surface of the remaining melt which frustrates further oxidation. Only where this seal is fractured by the force of the oxygen jet driving the melt out of the cut zone can the oxygen get access to unoxidised aluminium which then instantaneously oxidises to re-complete the oxide seal. In this way the oxidation process of aluminium is self limiting and the potential energy of the reaction is not freely available to the cutting process. These points will be amplified upon in the following sections.

The effectiveness of CO_2 laser cutting for any alloy is related to several complex and interrelated materials properties many of which change their value with temperature and phase changes:

Thermal conductivity
Melting point
Boiling point
Specific heat capacity
Effects of oxidation reactions (heat source, contamination etc.)

Latent heat of phase changes (e.g. solid–liquid)
Surface tension of melt generated
Material thickness
Reflectivity and surface finish of the material

These factors in combination determine the cutting speed (if any) possible for a particular laser–material interaction. Unfortunately the number of variables and their complexity make forecasting the behaviour of a material generally impractical. Trials are necessary whenever an unfamiliar material is to be cut.

All alloys which can be profiled at a particular laser power density are cut by melt shearing or a combination of this mechanism and exothermic oxidation. Melt shearing is a process whereby a localised molten pool is generated by the laser and simultaneously sheared by the cutting gas jet to produce a cut (see Fig. 3.1). A proportion of the liquid metal is ejected from the cut zone but, owing to the high surface tension of liquid metals, the gas jet cannot remove all the melt. The distinctive cut edge generated by this process is covered in rapidly solidified melt which often shows signs of turbulent flow during solidification. Another feature of such a cut is the presence of adherent resolidified melt or "dross" along the bottom of the cut edge (see Fig. 3.2). This dross is a result of the incomplete nature of the melt ejection under the action of the cutting jet. As the following

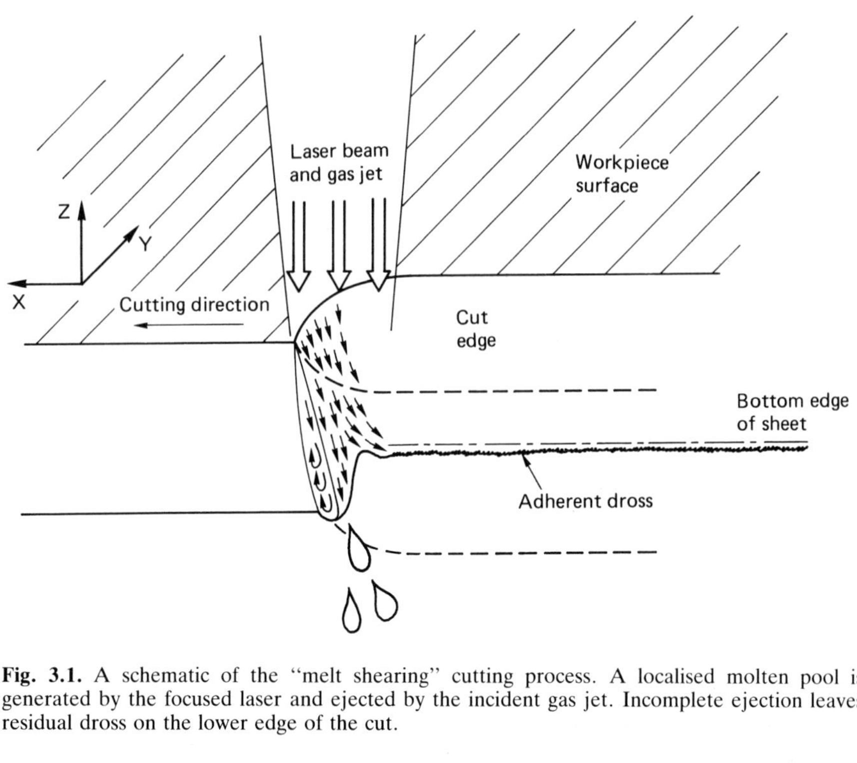

Fig. 3.1. A schematic of the "melt shearing" cutting process. A localised molten pool is generated by the focused laser and ejected by the incident gas jet. Incomplete ejection leaves residual dross on the lower edge of the cut.

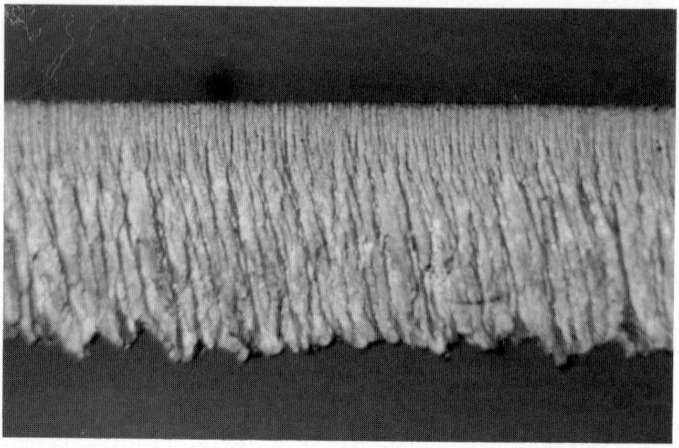

Fig. 3.2. An example of a cut edge produced in a sample of 4 mm (0.16 in) aluminium by the "melt shearing" process described in Fig. 3.1.

sections will reveal, it is possible to minimise dross by a number of methods, including laser pulsing, the use of very high cutting gas pressure or by an arrangement of secondary gas jets designed to manipulate the dross to the waste side of the cut.

The following sections outline the features of the cutting process for four major groups of commonly used alloys. The cutting speed results are a guide to the laser cutting performance of specific materials but these can be used to give approximate estimates for related alloys.

3.2 Titanium Alloys

The high strength to weight ratio and toughness of titanium alloys has ensured their increasing application particularly in high technology industries such as aerospace and medical equipment. The physical properties of these high cost alloys can give rise to problems when cutting them by traditional mechanical methods and laser cutting can provide a useful alternative. Carbon dioxide laser cutting of titanium alloys is divided into two techniques:

1. Laser–oxidation on cutting, which gives high cutting velocities but a product with an oxidised edge which generally needs to be subsequently removed. This process is useful for rough machining of components and selected applications where the oxidised edge does not affect component performance and service life.
2. Laser–inert gas cutting, which gives a high quality edge and can be used to produce finished components.

These two techniques will now be discussed separately.

3.2.1 Laser–Oxidation Cutting

3.2.1.1 General

This process relies upon laser initiated oxidation of the titanium in much the same way as steel cutting depends upon the oxidation of iron. The oxidation reaction is highly exothermic, generating a similar level of energy to that evolved by iron oxidation:

$$Ti + O_2 = TiO_2 + \text{Heat} \quad (912\,\text{kJ/mol at } 20\,°\text{C})$$

The laser acts to heat up the titanium sheet to a temperature at which it will ignite in the oxygen stream. The exothermic reaction which ensues is almost self sustaining, but requires an extra energy input from the laser in order to continue. In this way the laser movement dictates the line taken by the burning front and thus generates a cut. The process is extremely rapid and energetic if pure oxygen is used as the cutting gas and can be difficult to control if complex profiles are to be cut. The oxidation reaction can be inhibited and the process rendered more manageable if the oxygen is diluted with air or if air alone is used as the cutting gas. Inhibiting the oxidation process by using air or dilute oxygen reduces cutting speeds and the maximum sheet thickness which can be cut at a particular laser power (see Fig. 3.3). The cutting mechanism is changed from laser initiated oxidation to a combination of oxidation and melt shearing. This change in mechanism has the following effects on the cut quality:

1. Although the surface of the cut edge tends to be less rough, droplets of resolidified melt (dross) can be observed clinging to the lower edge of the cut. These droplets of dross need to be removed by a mechanical cleaning operation such as grinding or rumbling.
2. The extent of the oxygen contamination of the cut edge increases as does the heat affected zone. These effects are a result of the much

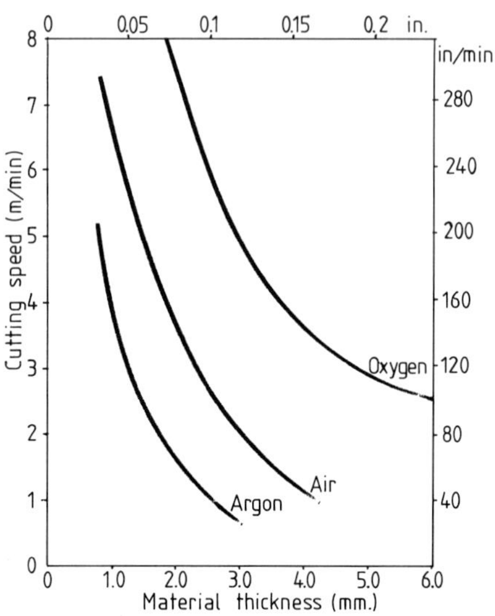

Fig. 3.3. Guideline cutting speeds for titanium alloys with a laser power of 1000 watts and a selection of cutting gasses (argon, air and oxygen). *Notes*: (1) If an unoxidised edge is required, argon must be used (3–10 bar) and the cut edge must be silver in appearance. Any blue or yellow colours indicate oxidation. High pressure cutting can result in a dross free edge. (2) Cutting with oxygen is potentially hazardous and can lead to an uncontrollable burning reaction, particularly when cutting thicker sections. The sparks are also highly energetic and can cause fires in the fume extraction system etc.

increased laser–material interaction time at any point along the cut, which is in turn a direct result of the lower cutting speed.

3.2.1.2 *The Effect of Oxygen on the Metallurgy and Mechanical Properties of Titanium*

Titanium alloys, like all other metals, are made up of a large number of microscopic crystals within which the positions of the various atoms are organised into a three-dimensional lattice. Wherever two or more crystals meet, the intermediate atoms assume compromise positions between the lattice positions which would be required from either of the crystals. The internal structure of the metal can therefore be seen to be made up of alternate zones of order and relative disorder. The ductility of titanium alloys relies upon the ease with which this order–disorder system can be deformed. The addition of oxygen to the system reduces the ease of deformation by introducing internal stresses. The oxygen diffuses into the material and occupies the gaps between the atoms in the crystals. (This situation is known as an interstitial solid solution.) The crystals have to expand slightly to accommodate the oxygen atoms and this strain gives rise to internal stresses. In this stressed state the crystals and their more disordered boundaries cannot reorganise themselves easily if the material as a whole is deformed. Therefore, as a result of the internal stress, the strength of the materials is increased but ductility is decreased.

At low levels of oxygen content this strengthening, hardening effect is welcomed and oxygen is added to some alloys to achieve the required mechanical properties. For example, the addition of oxygen to give levels of up to 0.2% in Ti-6Al-4V alloys is common. At levels in excess of these low values the strength of the material continues to increase, but this is accompanied by an increase in hardness and a decrease in ductility. Titanium alloys can dissolve oxygen up to levels in excess of 10%, at which point the material bears little relation to the ductile tough alloys in commercial use. During laser oxygen cutting the absorption of oxygen and other contaminants by the cut edge results in a highly stressed layer which is strong and hard but also brittle. During tensile loading, this layer can crack and the propagation of the crack will cause failure of the component. This effect is responsible for the poor fatigue life of components with heavily oxidised edges. The only way to recover the properties of the base material is to remove the oxidised edge by mechanical methods. This will generally involve the removal of between one and two times the thickness of the cut sheet.

3.2.2 Laser–Inert Gas Cutting

3.2.2.1 *General*

If titanium alloys are cut by melt shearing alone and protected from oxidation, the cut product retains much of the toughness and integrity of the base material. Some edge hardening is inevitable as laser cutting is a thermal process and there will always be a heat affected zone (HAZ) associated with the cut. This HAZ has far less effect on mechanical properties than oxygen contamination and in some cases can be disregarded, especially if the cut edge is to be subsequently welded (welding, of course, produces its own, larger, HAZ). Heat treatment can also be utilised to minimise any problem associated with HAZs. During inert gas cutting the major energy input to the cut zone as a result of exothermic reaction is absent. Although the absence of this reaction improves the physical and chemical quality of the cut, the cutting speeds are greatly reduced, compared with laser–oxygen cutting (see Fig. 3.3). Oxidation free cuts cannot necessarily be achieved simply by replacing the oxygen cutting gas with an inert gas such as argon or helium. The hot cut edges left behind after the passage of the laser have a strong affinity for atmospheric oxygen and nitrogen and will become oxidised with the associated blue discoloration. It is necessary therefore to keep these edges protected from the atmosphere until they become cool. One method to achieve this end would be to carry out the cutting in an inert gas chamber. This technique would give acceptable results but the use of such chambers is costly and inconvenient. A more direct method of protection can be afforded by use of a "dross jet" or auxiliary inert gas jets. These devices, which will be discussed in the next section, have been used with great success on titanium and other alloys.

3.2.2.2 The Use of a Dross Jet or Auxiliary Gas Jets

The use of a dross jet depends upon the principle that a laser cut generally has two sides of greatly different importance:

1. The component side of the cut
2. The waste material or scrap side

The aim of a dross jet or auxiliary gas jet is to propel all the dross issuing from the bottom of the cut zone onto the waste side of the cut. The principle of operation is shown in Fig. 3.4. If a stationary laser–moving workpiece cutting machine is being used a circular cluster of such nozzles can be arranged under the cut zone (Refs 34 and 37, Chap. 11). (This cluster must not move with the table.) Each nozzle is directed towards the cut zone and can be activated by computer control of its gas supply valve. During cutting, one or two jets may be employed at any one time to blow the dross towards the waste material side of the cut. For example, when cutting a square, the material movement sequence may be north, east, south and west. The gas jets employed would be those blowing east, south, west and north in that order (dual jets may be employed to keep the corners dross free).

These secondary jets under the cut zone act to deflect the main gas jet which acts coaxially with the laser beam as usual. It is this deflection which propels the dross to one side or another.

Although a device of this sort can be used to produce dross free components in any alloy which cuts by melt shearing, it achieves a double benefit when used with titanium alloys. It was mentioned earlier that a simple inert gas cutting jet acting coaxially with the laser beam may not sufficiently protect titanium from atmospheric oxidation. If an inert gas is

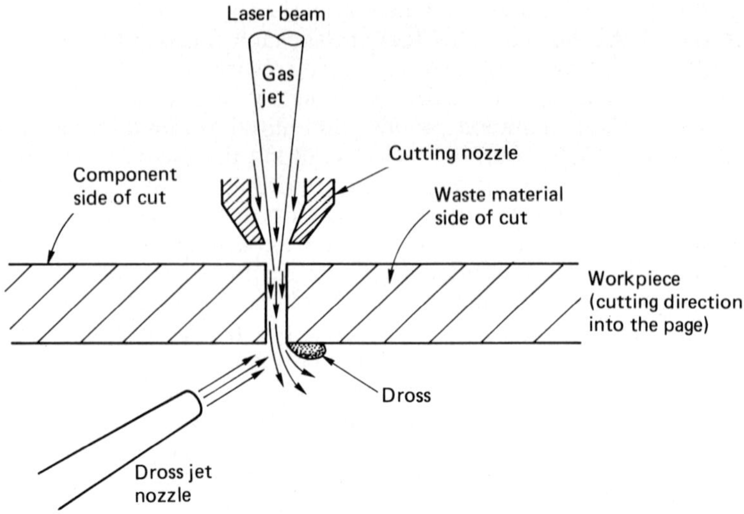

Fig. 3.4. The principle of operation of a dross jet designed to propel all the dross to the waste material side of the cut.

employed to pressurise the dross jet the level of contamination drops to acceptable levels as a result of two effects:

1. The additional inert gas shielding helps to keep the cooling cut edges free from oxygen and nitrogen.
2. The removal of all the dross to one side of the cut allows the dross free side to cool faster and therefore lowers the chance of contamination. Adherent dross on the lower edge of the cut acts as a "thermal reservoir" after the passage of the laser and keeps the cut edge at elevated temperatures for a substantial amount of time. Removal of the dross reduces the extent of the hot area behind the cutting zone minimising the chances of oxidation and making the area easier to shield.

These points are proven by the results given in Figs. 3.5 and 3.6. Figure 3.5 compares two cuts in thin section (1 mm) Ti-6Al-4V under identical conditions except for the use of a dross jet in the case of the edge shown in Fig. 3.5b. The absence of dross is obvious and the associated large reduction in oxygen contamination is shown by the graphs of Fig. 3.6. The lines on this graph compare edges cut by four methods:

1. Mechanically guillotined
2. Laser cut with a simple coaxial gas jet at a moderate pressure (argon, 1.1 bar through a 1.2 mm diameter nozzle)
3. Laser cut as 2 above with a higher pressure of 2 bar.
4. Laser cut using a dross jet in combination with the usual coaxial main jet

The measurements given in Fig. 3.6 were obtained by Auger electron spectroscopy. All the laser cuts were made at 1.8 m/min (71 in/min) with

a

b

Fig. 3.5. **a** A typical cut edge in 1 mm thick titanium alloy (Ti-6Al-4V) produced by the melt shearing mechanism described by Fig. 3.1. Note the droplets of dross on the lower edge of the cut (laser power 300 W; cutting speeds 1.8 m/min or 71 in/min; cutting gas argon). **b** The cut edge of a sample prepared under identical conditions to those of **a** except for the use of a dross jet which propelled all the dross onto the other side of the cut during processing.

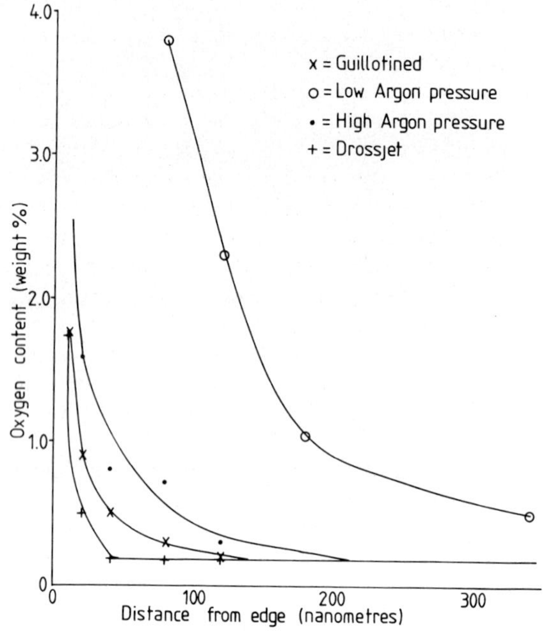

Fig. 3.6. Oxygen contamination levels on laser cut Ti alloy edges prepared by guillotining, laser cutting with a coaxial argon jet (1.1 and 2.0 bar) and laser cutting using the dross jet technique.

a laser power of 300 W, the same conditions as those used for the cuts in Fig. 3.5.

The guillotined edge is included here for comparison with the laser cut edges and represents an acceptable level of contamination. The two non-dross jet laser cuts show widely different contamination states both of which exceed this acceptable level. The higher pressure argon jet can be seen to give better results than its low pressure counterpart as a result of improved shielding of the cut zone combined with accelerated cooling by forced convection. The results obtained from the use of the dross jet show a marked improvement on these results and give contamination levels which are superior even to the guillotined edge. These improvements were noted over a very broad range of dross jet setting which implies that the supply pressure to the dross jet and the dross jet–cut zone distance are not critical. This lack of sensitivity can be explained by close observation of the bottom of a cut produced without the help of a dross jet. This reveals that the droplets are deposited evenly and alternately on one side of the cut and then the other. This is evidence of a delicate fluid dynamic balance which can be easily disrupted by a secondary gas jet blowing to one side across the bottom of the cut zone.

It is clear that the type of dross jet which uses a computer controlled cluster of nozzles under the cut zone is only suitable for machines where the cutting head is stationary. Even for this type of machine the fixing of the nozzles under the cutting head can be a problem as they must be attached to the floor by a cantilever of some sort which reaches under the workpiece. If batch production is being carried out, however, the CNC flexibility of the dross jet can be sacrificed and a much simpler system can be incorporated into the jig holding the workpiece.

Assuming the shape is to be profiled out of a metal sheet in a known location on the *X-Y* table, it is possible to construct a simple framework of copper tubing which lies to one side of the intended cut all along its length. Holes can be drilled along this tubing to generate a series of gas jets which, as they diverge and overlap, form a continuous flow of gas in one direction across the bottom of the entire cut. A schematic of such an arrangement is described in Fig. 3.7. In order to keep the gas flow as consistent as possible, it may be necessary to divide the copper tubing up into lengths, each of which has its own gas supply. The actual value of the gas flow across the bottom of the cut can vary substantially from place to place as only a small amount of lateral thrust is needed to upset the delicate fluid dynamic balance which would otherwise deposit dross equally on both sides of the cut.

This auxiliary gas jet arrangement is particularly appropriate to large three-dimensional, five- or six-axis cutting machines. Components profiled by these machines are always carefully located on jigs and the copper piping can be arranged as part of the jig. These large cutting machines are often used simply to trim the waste material off pressed components. In these cases the pierced copper tubing forms a ring around the perimeter of the jig, blowing the dross outwards onto the waste material.

Designing auxiliary dross manipulation jets for small, detailed work is much more difficult than for larger pieces. To take the simple example of a 3 cm (1.2 in) diameter disc; a ring of copper tubing would obviously be inappropriate. A different method of producing a radially outward gas flow could be to aim a pressurised gas jet perpendicularly at the centre of the circle or to enhance the radial flow from such an arrangement by using an inclined baffle. Other small pieces may require the fabrication of specially made platforms which sit under the cut component and have a peripheral ring of holes to allow the passage of the pressurised gas.

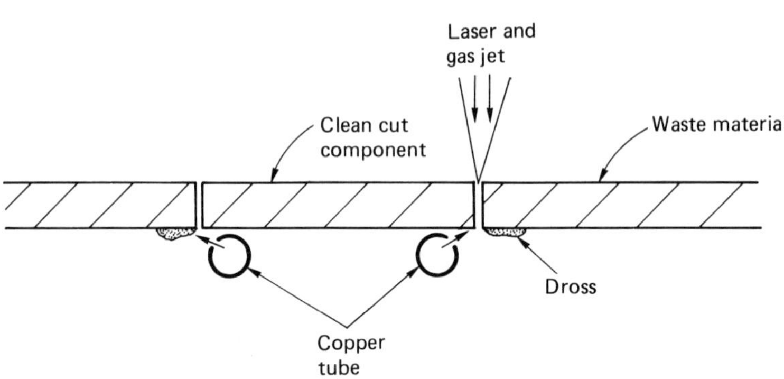

Fig. 3.7. A cross section of a copper tube dross jet designed for the production of a particular component. The tubing has holes every few millimetres (e.g. 6 mm (0.25 in)) to produce overlapping gas jets which propel the dross to the waste material side of the cut.

3.3 Aluminium Alloys

3.3.1 Introduction

Aluminium has three inherent properties which present a problem to the laser cutting process:

1. It has a very high reflectivity (low absorptivity) to CO_2 laser light.
2. It has a high thermal conductivity.
3. Its oxidation reaction is self extinguishing as it forms an oxide which is impermeable to oxygen.

Despite these drawbacks, aluminium alloys are commonly laser cut with great commercial success, although cutting speeds and the maximum thickness cut are lower than for steels (see Fig. 3.8).

It has been mentioned that all metals are highly reflective to the 10.6 μm radiation from CO_2 laser at room temperature. The small amount of light absorbed initially raises the temperature of the metal in the laser–material interaction zone and this improves the absorptivity, which in turn leads to further heating. In most metals this increased heating–increased absorption cycle is rapidly established and quickly generates a local molten pool of high absorptivity. In the case of aluminium the heating–absorption increase effect is difficult to establish, as the small amount of energy (~2%) absorbed

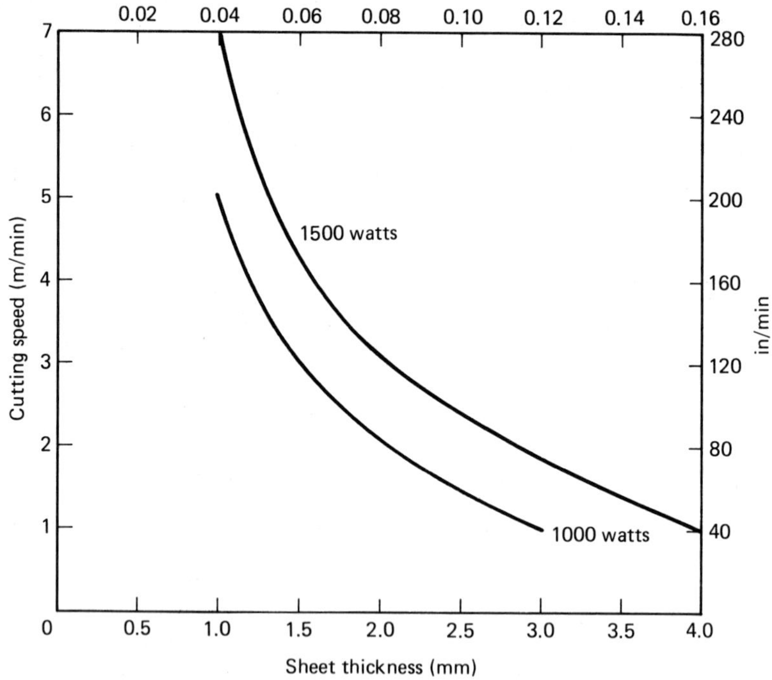

Fig. 3.8. Typical cutting speeds for aluminium alloys at laser powers of 1000 W and 1500 W. Cutting gas: oxygen at pressures of 2–6 bar. (*Note*: Anodised aluminium cuts approximately 30% faster.)

at ambient temperatures is rapidly dissipated away from the laser–material interaction area as a result of the high thermal conductivity. In this way the reflectivity and conductivity act together to frustrate the cutting process. A certain minimum energy density is required to melt and subsequently cut aluminium sheet of any thickness.

In practical terms, for continuous wave operation, a power of 500 W or more is required before aluminium cutting becomes commercially practicable. At lower powers the cutting speeds (if any) are very low, which of course means that the laser–material interaction time in any one area is relatively large. At these large interaction times the heat loss from the cut zone as a result of thermal conduction becomes considerable and below a certain laser power this will be large enough to prevent successful cutting. At higher powers and energy densities cutting speeds increase and conduction losses become less as the laser–material interaction time diminishes (see Chap. 10). It is this dependence of the level of heat dissipation on the cutting speed which renders the cutting process progressively less efficient as the sheet thickness is increased. To summarise this point, as cutting rates are increased the efficiency of the process also increases. As cutting rates are necessarily decreased at lower powers or increasing material thickness the process becomes progressively less efficient and eventually becomes impractical. (NB: Other thermal losses to the cutting process due to convective cooling from the gas jet and re-radiation of heat from the melt also increase with increasing laser–material interaction time. These losses, however, are very small compared with conductive losses when metals are considered, particularly high conductivity metals such as copper and aluminium.)

The loss of large amounts of energy from the cut zone as a result of the high thermal conductivity has an obvious side effect; the workpiece material as a whole becomes heated. Laser cutting depends on the localised nature of the cutting zone. This overall heating of the workpiece combined with the relatively low melting point of aluminium can change the thermodynamic equilibrium of the process and cause the molten zone to spread laterally. The cutting process can therefore become unstable, especially at corners, where the heat build up is greater as the laser–material manipulation equipment has to decelerate. Methods of dealing with this problem of workpiece overheating include laser pulsing, which will be dealt with later in this section, and the use of a coolant spray around the cut zone (see Chap. 2). Coolant (usually water) sprayed into the area around the cutting nozzle will be kept out of the actual cut zone by the high pressure oxygen jet but will be able to flow down the already cut edges. In this way the surplus heat in the workpiece is removed and the cutting zone remains localised.

In addition to the problems associated with high conductivity and reflectivity there is a basic inefficiency of the oxidation reaction which has already been referred to. Oxidation cutting is a process by which an oxide is exothermically produced on the surface of the molten metal. In some cases this surface effect can be self limiting or self extinguishing if the oxide generated remains firmly bonded to the underlying unoxidised material and is impermeable to the passage of oxygen. The oxide thus forms a seal on the metal surface which prevents further oxidation. The oxide produced during the cutting of aluminium alloys is the same as that which forms instantaneously on aluminium at ambient temperatures, i.e. Al_2O_3. A thin coating of this oxide covers all aluminium based materials and is the reason why aluminium is corrosion resistant. In fact aluminium itself is not at all corrosion resistant, but once its surface has instantaneously "corroded" to a thin layer of Al_2O_3 then no further corrosion is possible as oxidising (corroding) elements can not pass through this layer. The layer of Al_2O_3 is only 1–2 µm thick and is therefore effectively transparent, which is why we see the metallic surface below.

It is the protective nature of this oxide which frustrates the extremely exothermic oxidation reaction of the material. Cutting speeds are increased by only about 30% if oxygen is used as a cutting gas rather than an inert gas or air (cf. mild steel where the cutting speed with oxygen is approximately five times the speed achievable with air).

The level of oxidation which does take place during cutting is made possible by the repeated fracturing of the oxide film as a result of the turbulent flow of melt out of the cut zone. The dross left adhering to the lower edge of the cut is not divided up into droplets as in the case of titanium. Aluminium dross generation is a more continuous process which gives rise to the type of cut edge seen in Fig. 3.2. The liquid remaining on the cut edge after cutting has a high surface tension as a result of its rapid cooling by the underlying substrate and its surface layer of Al_2O_3. Aluminium oxide solidifies at temperatures of approximately 2017 °C and so can solidify on the surface of the, still molten, aluminium (melting point 659 °C) and thus increase its surface tension. The dross produced during aluminium cutting consists of aluminium coated in a thin layer of Al_2O_3 and therefore has much the same mechanical properties as the base material. It is easy to remove the dross by mechanical means (grinding, scraping etc.) or it can be manipulated to the waste material side of the cut by use of a dross jet (see previous section on titanium). The propellent gas from the dross jet in the case of aluminium does not need to act as a chemical shield (as in the case of Ti) and so air can be used.

3.3.2 The Effect of Surface Condition on Cutting (Anodising etc.)

Aluminium is used for a wide variety of applications from structural components to display. For most display work and for many engineering applications where mechanical suitability is combined with a high quality appearance (control panels etc.) aluminium is used in its anodised condition. Anodising is a method of increasing the depth of the oxide surface on aluminium by electrolytic methods. The aluminium sheet is immersed in a suitable electrically conductive liquid and becomes the anode in an electrical circuit. Under these conditions the oxide (Al_2O_3) "grows" on the metal surface as a porous layer which can have a variety of appearances depending on the electrical and chemical conditions in the bath (e.g. pulsed current or the addition of dyes to the liquid). The finished product can be a number of colours of which silver-white and black are the most common and generally has a matt or dull finish. This layer of aluminium oxide is commonly 20 µm or more thick and is not only decorative but also extremely hard and wear resistant. (Some "silver-white" grades of anodised aluminium are difficult to distinguish from non-anodised sheet but can be easily identified by a scratch test.)

The presence of this oxide film on the aluminium has a dramatic effect on the reflectivity of the material to CO_2 laser light. Arata and Miyamoto (Ref. 46, Chap. 11) have shown that as the thickness of the surface oxide layer is increased the absorptivity of the material grows from approximately 2% to 100%. This 100% absorptivity is associated with oxide depths in excess of about 4 μm and so commercially anodised grades with oxide depths of ~20 μm have a surface which is fully absorptive. This high absorptivity has three beneficial effects on the cutting process:

1. The material is easier to pierce prior to cutting (see next section).
2. Cutting speeds are increased by ~30%.
3. The cutting process becomes less sensitive to changes in the process parameters and is therefore rendered much more reliable (e.g. a change in nozzle–workpiece distance as a result of a non-flat sheet will not necessarily cause problems).

An increase in cutting speed by ~30% may seem small as the absorptivity of the surface has risen from ~2% to ~100% but it must be borne in mind that only the leading edge of the beam impinges upon the workpiece surface. The majority of the beam is incident on the molten oxidised material in the cut zone which has the same reflectivity whether the workpiece has been anodised or not.

At this point it is worth mentioning two disadvantages to the use of anodised aluminium:

1. Anodised aluminium costs considerably more than unanodised.
2. Anodised aluminium components cannot generally be bent to shape or welded with the same ease that normal aluminium can.

Except for jobs where the material must be bent or welded after cutting it can be cost effective to suggest to a potential customer that they use anodised sheet. Apart from the cutting advantages, the finished product will be stronger, harder, corrosion resistant and will have a more attractive surface finish.

The piercing and cutting problems associated with non-anodised aluminium can be minimised by lightly roughening the material surface with a fine grain emery paper (a hand held floor sander can be used). The matt finish generated will improve the piercing and cutting characteristics of the material and in most cases the surface finish is not important to the finished product. If abrasion cannot be used to increase surface absorptivity a surface coating of ceramic dust can be applied by painting the aluminium with certain types of abrasive cream domestic cleaning fluid. Many of these cleaning fluids are simply suspensions of Al_2O_3 particles which, when dry, leave a surface which will respond in a similar way to anodised sheet. Naturally the white deposit can be washed off after cutting to reveal the unaffected workpiece surface.

3.3.3 Pulsed Laser Cutting

When cutting high reflectivity and conductivity materials such as aluminium superior results are often obtained by using the laser in its pulsed mode. The high peak powers reached during the pulse help to increase the thermal gradient in the cut zone and thereby keep the surface molten. The cyclic nature of the energy input also helps to de-sensitise the process to small parameter changes and as fluctuations in gas pressure or focal position. (If a c.w. beam is used a steady state cutting reaction is established which can be very sensitive to such changes.) A pulse setting must be chosen which allows these benefits without substantially lowering the average input power. Pulses should overlap to a large extent and should therefore have an on:off ratio of 5:1 or more.

3.3.4 Piercing a High Reflectivity Metal

Although the penetration process for steels has been discussed earlier (Sect. 2.1.4) piercing of high reflectivity metals poses a specific problem. During the cutting process a localised area of high absorptivity exists within the cut zone, but before this can be established the material must be pierced. At the beginning of this drilling process the laser beam is interacting with the cold, flat surface of the workpiece which will have a much higher reflectivity than the eventual cut zone. In some cases the reflectivity of the material at this stage is so high that a cut zone can never be established. If this is the case some surface roughening or coating will be necessary in the piercing area. (This area will generally be in the waste material and will not affect the finished component.) In less severe cases the penetration process may take several seconds. If this is the problem the time taken will generally be greatly reduced if the laser beam or the material are moved very slowly during piercing (e.g. 0.01 m/min or 0.4 in/min). This movement has the effect of de-stabilising the molten pool generated by the laser and ejecting it to one side as the incident gas jet is decentralised.

Using the laser in its pulsed mode to penetrate high reflectivity materials will usually produce superior results to those obtained by a c.w. beam. The high peak powers in the pulses assist the coupling process to generate a melt and in some cases can be used to drill a hole by vaporisation. If vaporisation is the basic penetration mechanism (characterised by an intense blue light) then the "slow movement" technique described above will probably not be effective as it relies on fluid flow in the drilling zone.

Warning: Whenever high reflectivity metals are being pierced there is a danger of powerful stray reflections. These are generally most intense in the direction almost perpendicular to the surface of the workpiece and can easily damage the lens holder and cutting head etc. These reflections are also a possible hazard to personnel and safety precautions such as polycarbonate screens may be appropriate. Safety goggles should of course be worn at all times in any case when operating or watching a machine.

A material which cannot be pierced may still be cut satisfactorily if it is mechanically pierced in the cut start positions prior to cutting.

3.4 Nickel Alloys

Although nickel alloys are more difficult to laser cut than steels, they do not generally present as big a problem as the other alloy systems discussed in this chapter. Nickel alloys include a wide range of materials such as monels (nickel–copper); nimonics, inconels and hastelloys (nickel–chromium–iron); renes (nickel–chromium–cobalt–molybdenum) and of course nickel itself.

The discussion presented in this and previous chapters has emphasised the important influences the chemical composition of a metal has on the cutting process. The most important material properties which have a direct effect on cutting are reflectivity at a wavelength of 10.6 μm, thermal conductivity, melting point and the energy available from the oxidation reaction (if any). Nickel alloys include some alloying elements which accelerate the cutting process (e.g. iron) and others which tend to frustrate the interaction (e.g. copper). The reflectivity (at ambient temperature) of this range of alloys can vary from ~94% to 97% and thermal conductivities also vary considerably.

The exothermic reaction

$$Ni + \tfrac{1}{2}O_2 = NiO + heat \quad (244\,kJ/mol \text{ at } 20\,°C).$$

can be used to accelerate cutting speeds and oxidation cutting is particularly effective in the case of nickel–iron alloys, which cut very much like stainless steels. It must be borne in mind, however, that many of the complex nickel alloys used in aerospace applications (e.g. nimonics) possess a complex metallurgical structure which may not be suitable to thermal cutting techniques. The metallurgical effects of laser cutting are minimal in comparison with other thermal techniques (flame, plasma etc.), but in certain aggressive working environments (e.g. jet engines), even the narrow residual melt–HAZ region generated by this method can be found unacceptable. In most cases, however, laser cutting with all its advantages can be used on nickel alloys with excellent results. If effects such as preferential oxidation of certain alloying components or the presence of a hard oxide surface on the cut edge are a problem, higher quality cuts can be made using an inert gas cutting jet. Excellent results have been achieved by using high pressure (⩾10 bar) inert gas cutting jets which give a flat, smooth and oxide free cut edge with no adherent dross. The process is, however, more expensive than oxygen cutting for three reasons:

1. The high volume of inert gas consumed
2. The reduction in cutting speed by 50% or more (depending on material)
3. The need for a specially designed high pressure cutting head and appropriate (thicker) lenses

Although high pressure inert gas cutting gives qualitative advantages over laser–oxygen cutting it must be appreciated that a heat affected zone will always be a feature of a laser cut edge unless it is removed by subsequent heat treatment.

In general, most nickel alloys have much the same quality of cut as stainless steels. Cutting speeds will be between ~80% of the stainless steel values for some of the iron bearing inconels down to ~50% or less for copper bearing monels (see Fig. 2.20 for stainless steel values). The maximum thickness of the workpiece will also be reduced by the same order of magnitude depending on material chemistry.

3.5 Copper Alloys

With the exception of gold and silver, copper has the highest thermal conductivity and reflectivity to CO_2 laser light of any metal. It has been mentioned earlier that the combination of these two physical properties frustrates the establishment of a localised molten pool necessary to the laser cutting process. In spite of this, recent advances in the integrity of high power CO_2 laser beams have allowed copper and its alloys to be successfully cut at commercial rates in thicknesses of up to approximately 4 mm (0.16 in). Conductive thermal losses from the cut zone increase dramatically with increasing material thickness and this means that the most cost effective cutting is carried out on thin section sheet. The most commonly used copper alloys are bronze and brass, although bronze is rarely used in sheet form and is therefore not cut by laser. It is interesting to compare the thermal conductivities of these materials:

Pure copper	385 (W/m/K at 293 K)
Bronze (90 Cu–10 Sn)	180
Brass (70 Cu–30 Zn)	110

Although bronze can be disregarded as far as laser cutting is concerned, brass is often cut by this method. The zinc content of brass aids the cutting process in two other ways as well as reducing the thermal conductivity, compared with pure copper:

1. Although copper does evolve heat during oxidation the reaction is not as exothermic as it is for many other metals (e.g. iron, titanium, nickel etc.):

$$Cu + \tfrac{1}{2}O_2 = CuO + \text{heat} \quad [155\,\text{kJ/mol at } 20\,°\text{C } (293\,\text{K})]$$

The zinc oxidation reaction evolves approximately twice this amount of heat:

$$Zn + \tfrac{1}{2}O_2 = ZnO + \text{heat } [348\,\text{kJ/mol at } 20\,°\text{C } (293\,\text{K})]$$

This energy of reaction is combined with the reaction of copper to make the response of the material as a whole more vigorous. The extra thermal input to the cut zone helps to accelerate cutting rates.

2. The reflectivity of brass is lower than that of copper. The reflectivity of a metal at wavelengths above 2 μm is approximately proportional to the square root of its electrical resistivity. The electrical resistivity of brass is approximately four times that of copper and so the absorptivity is double.

Table 3.2 gives a number of examples of cutting speeds possible for copper and its alloys. For materials of such high reflectivity and conductivity there is a lower limit of laser power at which cutting can be carried out. Brass and copper cutting is usually only carried out at powers in excess of 1 kW.

Table 3.2. Laser–oxygen cutting speeds for copper alloys

	Thickness		Laser power	Cutting speed	
	in	mm	W	in/min	m/min
Copper	0.04	1	1500	120	3.0
Copper	0.08	2	1500	40	1.0
Copper	0.04	1	1000	60	1.5
Brass	0.04	1	1500	200	5.0
Brass	0.08	2	1500	80	2.0
Brass	0.04	1	1000	120	3.0

NB: Cutting speeds (if any) are highly dependent on surface condition and alloy content.

Laser–oxygen cutting of these alloys is generally far more successful than inert gas cutting. The reason for this is a combination of the extra heat generated in the cut zone and the improved absorption characteristics of the oxide layer which coats the cutting zone.

In many cases, particularly in the sign industry, copper or copper alloys are used for corrosion free display purposes. If a potential user needs a copper or golden coloured sheet for this type of application, the problems associated with copper alloys can be ignored if copper, brass or bronze plated stainless steel is cut instead. In this way complex display images with the correct appearance can be produced with a large saving in cost in materials and laser time.

3.6 General Comments

It is clear that the cutting of non-ferrous metals is generally more difficult than the cutting of steels. Many non-ferrous metals cut with similar speeds and edge qualities to stainless steels but speeds are greatly reduced if the thermal conductivity and reflectivity is high. Most commonly used metals and alloys are now cut by CO_2 laser on a commercial scale. Gold and silver are examples of materials which are not processed by CO_2 laser but applications do exist where these metals have been cut commercially by Nd:YAG lasers (see Chap. 8). Nd:YAG lasers are more suited to this type of high conductivity–reflectivity material as their shorter output wavelength of 1.06 μm is more efficiently absorbed than 10.6 μm CO_2 light.

The absorptivities (or reflectivities) referred to throughout this discussion are those of the solid metals at ambient temperatures. They are useful for comparison between metals, but do not give a true indication of the actual reflectivity of the cut zone. The actual reflectivity is a complex function of the following features of the cut zone:

1. Macroscopic geometry, i.e. the general inclination of the cut face and the lag between the leading edge of the laser beam and top of the cut zone.
2. Microscopic geometry, i.e. turbulence and ripples on the surface of the liquid.

3. Time based variations in 1 and 2.
4. The physical state of the material in the cut zone (liquids, gases and solids).
5. The chemical state of the material in the cut zone (e.g. oxidised etc.).
6. The temperature of the cut zone and its uniformity or otherwise.

Future work will undoubtedly reveal the actual reflectivities of cut zones for various materials but at the moment (1992) reflectivity comparisons can only be of a qualitative nature (see Chap. 10).

Although the preceding notes cover some of the various aspects of the cutting of non-ferrous metals, there is not yet enough information available for the generation of an accurate library of processing parameters. This being the case, the only way to identify accurately the cutting speeds and qualities possible for a particular laser–alloy content is to have actual trials carried out.

Chapter 4

Cutting Non-metals

4.1 Introduction

Most non-metallic materials are highly absorptive at the CO_2 laser wavelength of 10.6 μm. For this reason they are cut very efficiently and it is not surprising that one of the first ever (1971) commercial uses of CO_2 lasers was the cutting of plywood die boards for the packaging industry.

The cutting process involves three mechanisms of which one is usually dominant for any particular non-metallic material:

Melt shearing
Vaporisation
Chemical degradation

These cutting mechanisms determine the cut edge quality possible in each case and will therefore be extremely important to the following discussion. As an introduction the three mechanisms can be defined as follows.

4.1.1 Melt Shearing

Melt shearing is a term which can be used to describe the situation where the laser generates a small molten pool which is continuously blown away by the gas (air) jet acting coaxially with the beam (see Fig. 4.1). It is important to realise that, in principle, the material which is ejected during cutting has the same chemistry as the material being cut and has therefore experienced only a physical phase change from solid to liquid. In practice the resolidified ejected material will generally have undergone a certain amount of chemical change (depolymerisation etc.). Materials which are cut by melt shearing (e.g. most thermoplastics) are cut very quickly with a high quality edge which may be covered in microscopic ripples associated with the fluid dynamics of melt ejection from the cut zone.

4.1.2 Vaporisation

Polymethyl methacrylate, which is more commonly known as acrylic, perspex or plexiglass etc., is the only commonly used material which is laser cut by vaporisation. The acrylic boils to give off a vapour of methyl methacrylate

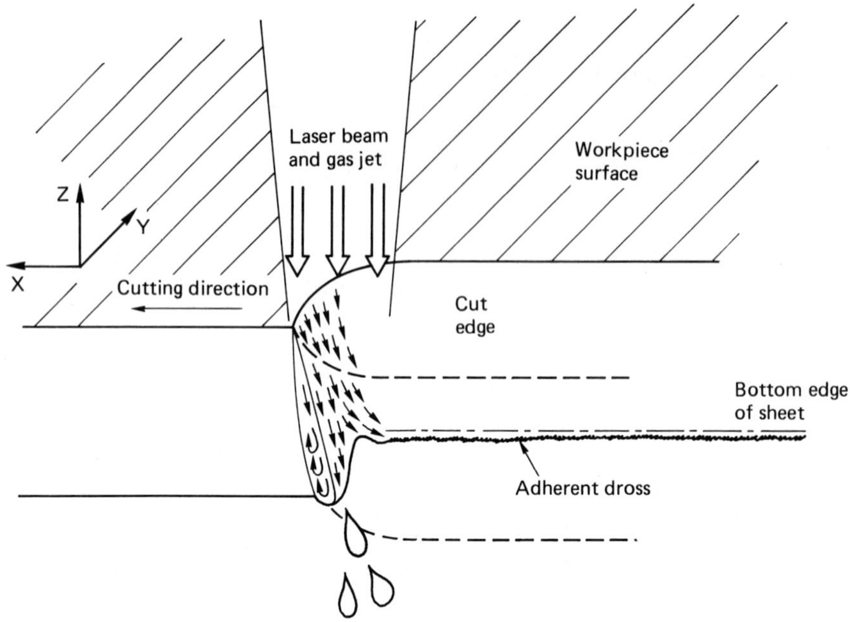

Fig. 4.1. A schematic of the melt shearing mechanism by which most thermoplastic polymers are cut. The laser locally heats and melts the material which is subsequently blown out of the cut zone by the air jet.

and so it can be appreciated that the minimum amount of chemical degradation is carried out and the process approximates to a purely physical phase change from solid to liquid to vapour. The cut edge is of extremely high quality and superior to a flame polished edge. Another polymer which is cut by this mechanism is polyacetal.

4.1.3 Chemical Degradation

Chemical degradation, as the name suggests, involves the laser carrying out a great deal of chemical work during cutting. The energy from the laser beam acts to break chemical bonds and thus disrupt the integrity of the material. When cutting a material such as wood this degradation process has the effect of reducing the large cellulose molecules down to their constituent parts of carbon and water vapour etc. Cutting by chemical degradation tends to be a slow, high temperature process, compared with melt shearing or vaporisation, both of which involve simple physical phase changes. Cut edges generated by chemical degradation tend to be flat and smooth but covered in a fine layer of residual carbon dust.

This chapter will discuss the cutting of non-metallic materials under the following headings:

Polymers
Wood based products

Ceramics
Composites and miscellaneous materials

4.2 Polymers

It is convenient to segregate polymers into three groups according to the dominant material removal mechanism which generates the cut. As an aid to the reader, the following is a list of the most commonly utilised materials in each group:

1. *Melt shearing*. Materials include most thermoplastics, e.g. polypropylene, polystyrene, polyethylene, polyamide (nylon) and acrylonitrile-butadiene-styrene (ABS).
2. *Vaporisation*. Polymethyl methacrylate (acrylic) and polyacetal.
3. *Chemical degradation*. Polymers include most thermosets, such as phenolic or epoxy resins and rubber products. Poly vinyl chloride (PVC) is also cut by this mechanism, giving off highly toxic fumes of HCl. Other important materials include polycarbonate and polyurethane which also gives off toxic fumes (HCN).

4.2.1 Cutting by Melt Shearing

4.2.1.1 Cut Quality

As mentioned earlier, cutting by melt shearing involves the ejection of a laser generated melt from the cut zone. This ejection is carried out by the air jet which acts coaxially with the focused laser beam. Depending on the cutting conditions and the surface tension of the liquid polymer, the material leaves the cut zone as droplets and/or pressure extruded fine threads. These fine threads or filaments may initially remain attached to the lower edge of the cut, but are generally broken off under the action of the pressurised air jet leaving the cut zone. In some cases residual filaments may need removal from the lower edge of the cut by scraping, but this is a very simple operation.

It is not possible for a gas jet to remove all the liquid from the cutting zone, and a layer is always left on the surface of the cut edge. Although the cut edge is of very high quality, the presence of resolidified melt is responsible for two features, both demonstrated in Fig. 4.2.

1. The cut surface is generally covered in microscopic ripples as a result of the residual melt layer experiencing the turbulence of the gas jet during solidification.
2. A small amount of resolidified melt is left attached to the lower edge of the cut. This "dross" can be easily removed by scraping or abrasion, but this is not usually necessary.

In general the laser cut edge of a polymer cut by melt shearing is superior to a mechanically cut edge. Kerf widths are small (0.1–0.6 mm or 0.004–

Fig. 4.2. A typical melt-sheared cut edge on a sample of 4 mm thick ABS. Note the microscopic ripples on the cut edge and the small amount of adherent dross on the lower edge of the cut.

0.02 in) and complex shapes can be cut as the minimum detail radius possible is equal to the kerf width. In many cases the difference in edge quality between a laser cut and an injection moulded sample is minor and for this reason laser cutting can be used to cut a number of chosen profiles out of a standard injection moulding. This approach has been used, for example, to produce a wide range of computer keyboard arrangements from a mass produced ABS blank.

4.2.1.2 The Effects of Process Parameters on Cut Quality

Cutting Speed. If the cutting speed is set too low for a particular laser–material combination, the effects on the cut edge quality will be negligible except in extreme cases where the major problem will be the reflection of unused energy off the *XY* table onto the bottom of the sheet. This reflected light can damage the polymer by melting or by stress relieving the sheet, which will result in distortion of the finished component. Most extruded, rolled or moulded polymer sheet has a certain amount of residual stress and will irreversibly warp if heated above a critical temperature. On the subject of thermal distortion, many polymers combine a low thermal conductivity with a high coefficient of expansion, and so general heating of the sheet during cutting should be avoided if an accurately cut, repeatable shape is required.

If the cutting speed is set too high the cut will not fully penetrate the material and a groove will be produced in the material surface. As the melted material cannot leave the laser–material interaction zone in the usual way (i.e. out of the bottom of the cut), this groove is surrounded by resolidified melt on each side.

Cutting Gas Pressure. If the cutting gas pressure is too low there is a danger of the laser igniting the stream of droplets leaving the cut zone. The heat from any such flame can distort, stress relieve and even ignite the work piece. A flame of this type is shown in Fig. 4.3. Another feature of in-

Fig. 4.3. Ignition of the stream of material ejected from the cut zone can be a problem when cutting non-metals. In this case the vapour ejected when cutting acrylic sheet has ignited to give a yellow sooty flame.

sufficient gas flow through the cut zone is a reduction in achievable cutting speed as the molten material is ejected less effectively.

If the gas pressure is excessive, the increased turbulence in the cut zone can give rise to an increase in surface roughness of the cut edge. Cutting speeds can be affected as too high a flow rate of gas will tend to refrigerate the cut zone and render the cutting process less efficient. The process also tends to generate filaments rather than droplets at higher air flows. This can be a nuisance as extraction systems can become blocked.

Optical Considerations. The focal length of the cutting lens is related directly to the focused spot diameter. A reduction in focal length will give a smaller focal spot diameter and therefore a higher energy density. This will, in general, mean an increase in cutting speeds and a reduction in kerf widths. These benefits may, however, be outweighed by a reduction in flatness and perpendicularity of the cuts as a result of the lower depth of focus of the smaller focal length lens. This smaller depth of focus will also make the process more susceptible to disruption if the sheet is not flat.

The mode of the laser is only important to polymer cutting because it has an effect on the "focusability" of the beam and therefore the focused spot size. Anomalies in the spatial distribution of the beam's energy will not generally be apparent in the condition of the cut edge, because these materials are highly absorptive and melt at relatively low temperatures. The fluid flow in the melt zone integrates out any peculiarities in the mode. For this reason lasers designed to be used for polymers will not necessarily be of a high enough quality to cut "mode sensitive" materials such as mild steel (see Chap. 2). It is also worth noting that the effective focal depth for good quality cutting is much larger for polymers than it is for metals. At moderate cutting speeds (~75% of max) the effective focal depth can be as much as

±3% of the focal length of the lens (i.e. a 100 mm focal length lens will operate with the material at any distance of between ~97 and ~103 mm). This is a useful phenomenon if curved or buckled surfaces are to be cut.

4.2.2 Cutting by Vaporisation

4.2.2.1 Acrylic Sheet

Acrylic sheet is a material which has been one of the mainstays of the laser cutting industry ever since sign and display manufacturers discovered the excellent cut quality possible by vaporisation. It is the only material commonly cut by this mechanism.

Acrylic sheet, more commonly known as perspex or plexiglass, is available either as a colourless transparent sheet (pure polymethyl methacrylate or PMMA) or in a variety of colours in grades from transparent to opaque. Coloured sheets are tinted by organic dyes in the transparent types and by the addition of a coloured filler in the translucent and opaque grades. (Fillers can include a number of materials including dyed wood flour). All grades of acrylic can be laser cut, including preprinted panels. The inks used in the printing are also polymeric and do not interfere with the cutting process. The finished product is greatly superior to mechanically cut preprinted panels which tend to become delaminated along the edge.

During the cutting process the polymethyl methacrylate is melted and then dissociated into a vapour of methyl methacrylate, a colourless gas with a distinctive smell. This gas is highly absorptive of the laser light and can become ignited by the beam to give a yellow sooty flame (see Fig. 4.3). This flame should be avoided as it damages the acrylic and is best prevented by ensuring that the velocity of the gas cloud leaving the cut zone is greater than the velocity of the flame burning front which is then effectively "blown out". Two methods exist of increasing the velocity of the gas from the cut front: (1) increasing the cutting speed and (2) increasing the cutting gas jet pressure. The maximum cutting gas pressure is limited when cutting acrylic if the glossy flame "polished" edge associated with laser cutting is required (see Fig. 4.4). To achieve this very high quality edge the residual melt remaining on the cut edge must be allowed to resolidify without being disrupted by the incident air jet. If the air flow through the cut zone is increased above a certain (low) value the solidifying layer becomes turbulent and the finished cut edge has a frosted appearance (see Fig. 4.5). It is obvious therefore that when cutting display acrylic the air flow through the cut zone should be low enough to avoid frosting of the cut edge, but high enough to prevent ignition of the gas leaving the cut zone. In order to achieve the correct air flow it is recommended that a larger diameter nozzle than those used for metals is employed. A 2 or 3 mm (0.08–0.12 in) exit diameter nozzle allows the operator to select the correct gas flow levels within a larger operating window than that allowed by the more usual ~1 mm (~0.04 in) type.

A flat "polished" edge quality can be achieved for all clear or coloured transparent and translucent grades of acrylic. As the proportion of filler is increased to produce opaque or semi-opaque grades the cutting mechanism

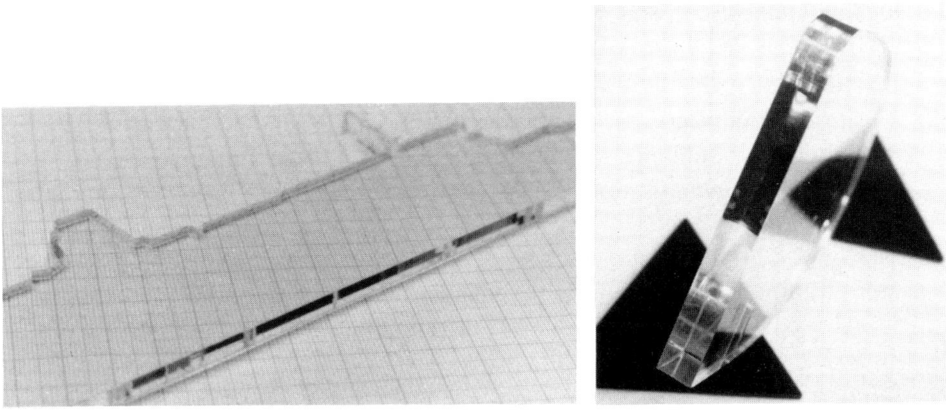

Fig. 4.4. Examples of the "polished" edge which can be produced when laser cutting acrylic with a low pressure air jet.

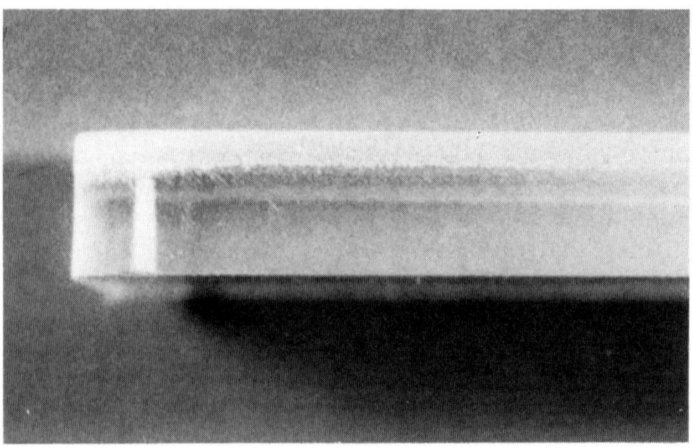

Fig. 4.5. The frosted appearance of a laser cut acrylić edge produced using a moderate or high pressure air jet.

changes to give a different cut quality. The result of cutting what is, in effect, a polymethyl methacrylate–filler composite is a reduction in cutting speed of up to 20% and a cut edge which is flat and smooth, but no longer polished in appearance. In the densest white grades of acrylic the filler degrades to give a darkened edge which can be removed by abrasion if necessary.

Acrylic sheet is often supplied wrapped in paper or polythene and can be cut in this wrapped state. If the wrapping is loose it is often better to remove it as hot vapour can become trapped between the wrapping and the sheet causing damage to the sheet surface. Also, loose wrapping can become ignited by the laser.

Although most acrylic is used for non-load-bearing display purposes, it is important to bear in mind that the "polished" surface of the laser cut edge can be prone to cracking which could cause failure in any mechanical loading application. This weakness of the cut edge is due to a combination of

1. Residual stress on resolidification.
2. Production of microscopic trapped pores in the solidified melt.
3. Directional solidification of the melt (perpendicular to the edge).
4. Local depolymerisation which reduces the molecular weight of the cut edge which in turn lowers the tensile strength.

The process of laser cutting has, however, a less severe effect than the more traditional process of flame polishing. In most cases this reduction in strength and increased tendency to cracking do not affect the service life of an acrylic component but where cheaper, lower molecular weight, extruded grades of acrylic are employed as display shelving etc., laser cutting might give rise to premature failure. Cutting the material with a higher air jet pressure reduces the tendency for edge cracking but, of course, produces a cut with a frosted edge. Problems can be alleviated by putting a radius on all internal corners but even this precaution may not help if the component is to be bent to shape before use. Trial samples should be cut, made up and tested whenever laser cut extruded acrylic is to be used in load bearing applications.

One other area where a "polished" edge can give rise to problems is if the acrylic is to be printed after cutting. The inks contain acrylic solvents which assist the binding of the colour to the material but can attack the cut edge. This chemical attack can lead to edge cracking which greatly weakens the component and looks unsightly. (Annealing the cut components will improve the tensile/fatigue strength and "printability" of the material if a "polished" edge is required. A typical annealing treatment is to bake the components at 80 °C for one hour.) The problems are minimised once again, by using high quality cast acrylic and cutting with high pressure air jet to achieve a frosted finish. As in the case of load bearing applications it is advised that sample trials should be carried out to identify any potential problems.

4.2.2.2 Polyacetal

Polyacetal sheet is less common than acrylic, but it has been cut with great success by CO_2 laser. This material can be used as a dry lubricant spacer and thus needs to be cut to shape to lie between moving metal components. During cutting the polyacetal generates large volumes of gaseous formaldehyde, which has a distinctive smell, but is not toxic in small quantities. The cut edge is flat, smooth, and has a polished appearance similar to acrylic cuts.

4.2.3 Cutting by Chemical Degradation

4.2.3.1 Thermoset Polymers

Thermoset polymers, which are cut by chemical degradation as the main material removal process, tend to generate a carbon based smoke during cutting and this is often accompanied by a residual carbon deposit on the cut edge. The majority of this carbon deposit can be removed by wiping, but a darkened cut edge is typical. The source of the carbon is obviously the material itself. Although thermoplastic materials such as polystyrene are carbon based, they do not degrade down to carbon during the cutting process because there is no need to do so. The monomers which join together to make up a thermoplastic polymer molecule have only two linking groups and form themselves into long chains. The bulk plastic material is made up of these long chains tangled together but not actually chemically joined. When the laser heats up an area during the cutting process, the material melts but the individual chains remain intact and are blown away from the cut zone as integral parts of the droplets formed. (The process can be likened to forking a channel across a large plate of spaghetti!)

In the case of thermoset plastics, however, the monomers which form the building blocks of the polymer have more than two linking groups and so, instead of forming linear chains of molecules, they join together as a three-dimensional matrix. In this case the laser cannot soften the structure and eject a certain number of polymer molecules in droplets because the molecules are chemically linked in all directions. It is necessary, therefore, to carry out chemical as well as physical work. The bonds which make up the polymer matrix must be destroyed. This chemical degradation takes up a lot more energy than the simple physical phase change of melting and it is not surprising that cutting speeds are much reduced and cut zone temperatures are a great deal higher than those experienced in melt shearing cutting (e.g. 3 mm phenolic resin: 2.9 m/min at 400 W, temp. ~3000 °C; polystyrene: 4.2 m/min, temp. ~1000 °C.)

Cut edges are generally extremely flat and smooth as a result of the non-fluid material removal mechanism (see Fig. 4.6). These cut edges compare well with mechanically cut specimens as thermoset materials are often hard and brittle, and thus prone to edge damage during cutting. Fume extraction is necessary as in all cases of polymer cutting because the fumes given off are inflammable. If the fumes are allowed to build up and collect they can suddenly (and explosively) ignite.

Fig. 4.6. A typical thermoset polymer cut edge (3.0 mm, 0.12 in thick Tufnol). The edge is flat, smooth and covered in a thin layer of black carbon dust (most of which can be removed by wiping).

4.2.3.2 Thermoplastic Polymers and Rubbers

Certain thermoplastic polymers are cut by chemical degradation. The most important materials in this group being poly vinyl chloride (PVC), polyurethane, and polycarbonate. The reason for the degradation rather than a melt shearing mechanism is the proximity of the melting points of the materials to their dissociation or degradation temperature. The fumes given off by PVC during laser cutting consists to a large extent of hydrogen chloride (HCl) which is highly corrosive and readily combines with any moisture to become concentrated hydrochloric acid. Other degradation products (Ref. 39, Chap. 11) include benzene, vinyl chloride, propene and toluene. Most commercially used PVC grades are less than 70% pure as plasticisers are used extensively to increase flexibility, but this merely dilutes the major problem of laser cutting PVC which is the generation of large volumes of toxic corrosive HCl. One method of dealing with this problem (Ref. 43, Chap. 11) is to incorporate gaseous ammonia into the air jet which removes material from the cut zone. The ammonia (NH_3) reacts with the hydrogen chloride to generate ammonium chloride, a non-toxic dust which can be fume extracted and/or dissolved in water by passage through a sprinkler system. Although this is technically a rather elegant solution to the problem, it is still intrinsically unsafe as two toxic gases are being played off against each other. The only safe way to deal with PVC cutting requests is to persuade the customer to change to another material. Personal experience has shown that PVC is often chosen because it is familiar and cheap. The cut edge of PVC is flat, smooth and covered in a dark brown dusty deposit.

The situation with polyurethane is not so clearly defined or so simply dealt with. To date (1991) no detailed analysis of the fumes generated by polyurethane during laser cutting has been carried, out, but there is a substantial risk of the generation of hydrogen cyanide which is, of course, potentially lethal.

Polycarbonate is cut by a combination of melt shearing and chemical degradation and has a distinctive cut edge quality. This edge combines the ripples associated with melt shearing and a slight discoloration due to the degradation. When laser cut, clear grades of polycarbonate have a yellowish edge discoloration. The yellow-white fumes given off during cutting have not yet been analysed, but are certainly unpleasant and could be toxic. Once more, care must be taken about fume extraction. Polycarbonate is often used as a shield material around CO_2 laser work stations as it combines transparency with a good resistance to ablation by unfocused beams.

Rubber materials are primarily made up of a monomer called butadiene which forms linear chain molecules (polyisoprene) similar to those mentioned in the discussion of thermoplastics. Commercial grades of rubber are generally much stronger than polyisoprene because sulphur has been added to form bridges between the individual chains. This cross linking (traditionally known as vulcanisation) can be tuned, depending on the amount of sulphur added, to give stiff or flexible rubber grades. The laser cuts sulphur cross linked polyisoprene by the chemical degradation mechanism generating a dense sooty smoke with a very unpleasant smell. The surface of the cut is usually covered in a sticky liquid of depolimerised butadiene which will eventually resolidify as the polymer, but which can be removed by wiping.

Complex shapes can be cut which would be difficult to produce by any other method as the rubber sheet obviously distorts under the influence of any mechanical cutting device. Supporting the flexible sheet during the laser cutting operation can be a problem but this can be overcome by cutting the rubber and an underlying sheet of a stiffer material such as acrylic, together. For applications where this approach would be too expensive in acrylic wastage, specialised jigging techniques as described in Sect. 5.5 can be employed.

Carbon powder is often mixed into the rubber material to change its physical characteristics which is why most rubber products are black in appearance. The presence of carbon in rubber sheet slows down the cutting process as carbon does not melt and degrades only at high temperatures. Carbon also raises the thermal conductivity of the material which tends to waste a proportion of the laser energy by conducting heat away from the laser–material interaction area. (This heat wastage is small compared with metal cutting.) The carbon is rarely uniformly distributed within the rubber sheet, particularly in thick sections, and this can lead to intermittent cut failure. For this reason the working cutting speed must be low, compared with the maximum speed for the uniform material. Carbon contents also change from batch to batch of the same grade of rubber and, of course, from grade to grade. The cut edge of high carbon content rubbers is generally smooth, flat and covered in a layer of carbon dust. A few typical cutting speeds for rubber and carbon filled rubber are given in Table 4.2 in the next section.

4.2.4 Cutting Speeds for Polymers

Commercial laser cutting dates from the beginning of the 1970s and it is therefore possible to accumulate a large number of isolated results referring to the cutting speed attained at a certain laser power for a particular material at a given thickness. For many materials insufficient data exists to present cutting speeds in a graphical form, and for this reason the technical literature has tended to rely on long lists of unrelated results. The information usually required by a potential laser user is concerned with the maximum cutting speeds achievable at various thicknesses for a particular material. Owing to the diversity of laser design, material manipulation systems and cutting head configurations (particularly focusing lens characteristics), accurate information is only made available by having trials carried out. This is particularly true where the material characteristics are not reliable (e.g. wood or hand laid glass fibre reinforced resin).

A book of this sort would, of course, be incomplete without information relating to cutting speeds at various material thickness and laser powers. The tables included in this section are intended to satisfy this demand, but must be treated with caution. The reasons why these results cannot simply be accepted as the exact truth are manyfold and include the following:

1. The majority of commonly used polymer names are not specific to a precise chemistry and refer to a group of materials which can have widely different properties. A good example is ABS which refers to any

Table 4.1. Typical cutting speeds for selected polymers using a 500 W CO_2 laser

Thickness		Acrylic (PMMA)		Polyethylene		Polypropylene		Polystyrene		Nylon		ABS		Polycarbonate		PVC	
mm	in	m/min	in/min	m/min	in/min	m/min	in/min	m/min	in/min	m/min	in/min	m/min	in/min	m/min	in/min	m/min	in/min
1	0.04	35.0	1400	11.0	430	17.0	670	19.0	750	20.0	780	21.0	830	21.0	830	28.0	1100
2	0.08	15.0	600	4.0	160	7.0	275	7.4	290	8.0	315	8.2	320	8.2	320	11.0	430
3	0.12	8.0	320	2.2	87	4.0	160	4.2	165	4.8	190	5.0	200	5.0	200	6.4	250
4	0.16	5.5	220	1.5	60	2.8	110	3.0	120	3.5	140	3.6	140	3.6	140	4.3	170
5	0.20	4.5	180	1.2	47	2.0	79	2.3	90	2.6	100	2.7	105	2.7	105	3.2	125
6	0.24	3.5	140	1.0	39	1.6	63	1.8	71	2.0	79	2.1	83	2.1	83	2.5	98
7	0.28	3.0	120	0.8	31	1.3	51	1.6	63	1.6	63	1.7	67	1.7	67	2.0	79
8	0.31	2.3	90	0.6	24	1.1	43	1.2	47	1.2	47	1.3	51	1.3	51	1.7	67
9	0.35	1.9	76	0.5	20	0.9	35	1.0	39	1.0	39	1.1	43	1.1	43	1.4	55
10	0.39	1.5	60	0.4	16	0.7	27	0.9	35	0.8	31	0.9	35	0.9	35	1.2	47
12	0.47	1.2	48	0.3	12	0.4	16	0.7	27	0.5	20	0.6	24	0.6	24	1.0	39

Notes: Cutting speeds can be changed dramatically by changes in molecular weight, degrees of crystallinity, and porosity.

As a first approximation, cutting speeds and maximum material thickness can be assumed to vary in a linear manner with laser power (between 100 and 1500 W).

The cutting gas is usually air at moderate pressures (1–4 bar). If a glossy edge is required on acrylic this may be dropped to below 0.25 bar. Nozzle diameters: 1–2 mm (0.04–0.08 in).

polymeric combination of three monomers: acrylonitrile (A), butadiene (B), and styrene (S). The strength, toughness and scratch resistance of an ABS material can be varied by altering the proportions of A, B and S, and this change in chemistry will affect the maximum cutting speed. (An even more variable material is wood, where cutting speeds can vary by an order of magnitude depending on density and water content.)

2. Cutting speeds are directly related to the energy density in the cut zone. This is affected by the laser power and the size of the focused spot. Although guidelines of cutting speed against material thickness are given in Tables 4.1 and 4.2, no information is provided about the focused spot size. If a shorter focal length of lens is used, the spot size decreases giving an increase in both energy density and maximum cutting speed. The mode of the laser beam also affects the ability of the lens to produce a small focused spot, any variation away from the TEM_{00} Gaussian cross section tends to enlarge the focused spot diameter and thus reduce cutting speeds.

For these reasons the cutting speeds given in Tables 4.1 and 4.2 can be accepted only as guidelines. For accurate data to be established, actual trials need to be carried out. These trials need not be extensive as operating curves can be produced from a minimum of information, e.g. cutting speeds at one power for three material thicknesses. Notes on how to produce your own cutting speed curves will be given in the two following sections. The first section describes a graphical method and the second explains the thinking behind the graphical method in terms of a simple formula.

Table 4.2. Cutting speeds for selected thermoset plastics, rubbers and fibre-reinforced materials with a CO_2 laser

Material	Thickness		Cutting speed		Laser power
	mm	in	m/min	in/min	W
Formica	1.6	0.06	7.8	312.0	400
	1.6	0.06	14.0	560.0	1200
	3.0	0.12	2.8	112.0	400
Phenolic resin	3.0	0.12	2.9	116.0	400
	6.0	0.24	1.1	44.0	400
Rubber	3.0	0.12	4.0	160	400
	6.0	0.24	1.6	64	400
	9.0	0.35	0.9	36	400
	12.0	0.47	0.6	24	400
Rubber (carbon filled, black)	3.0	0.12	3.0	120	400
	6.0	0.24	1.2	48	400
	9.0	0.35	0.7	28	400
	12.0	0.47	0.4	16	400
Fibreglass (glass-reinforced epoxy resin)	1.6	0.06	5.2	208.0	450
	1.6	0.06	15.0	600.0	1200
	3.2	0.13	2.4	96.0	400
Glass filled nylon	3.0	0.12	2.6	104	400

Notes: Cutting speeds for materials such as fibreglass depend on the relative proportion of glass, resin, and trapped air in the material.

Cutting gas is high pressure air (3–10 bar) nozzle diameters: 1–2 mm (0.04–0.08 in).

Cutting speeds will increase dramatically if porous grades of rubber are cut – figures given here are for fully dense material.

4.2.5 Cut Speed Forecasting for Polymers Using a Standard Processing Curve

4.2.5.1 The Graphical Method

The graphical method of producing a cutting speed versus material thickness curve relies on two factors which can be assumed to be true for the cutting of polymers:

1. The maximum cutting speed is proportional to the laser power (assuming constant focused spot size), i.e. twice the power gives twice the cutting speed. (This remains true only if the optical arrangement remains constant, a change of laser or lens focal length will change the focused spot size.)

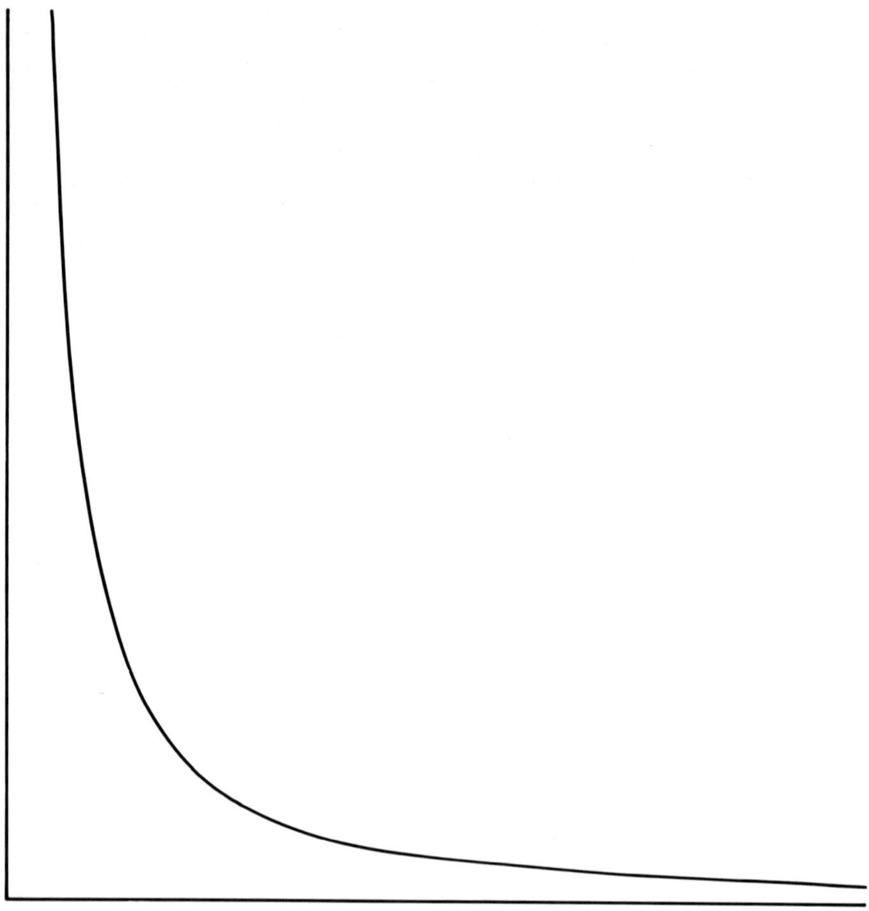

Fig. 4.7. This unmarked graph gives the correct approximate shape for the cutting speed curve for any polymer. This graph can be photocopied and used in the way described in Sect. 4.2.5 and Fig. 4.8 to produce a full cut speed curve from only one or two practical trials.

2. The shape of the cutting speed curve is approximately the same for all polymers, although the actual values given on the axes of the graphs changes. (A comparison of the information given in Table 4.1 will reveal the similarity in cutting curve shape.)

The curve given in Fig. 4.7 is the correct approximate shape for any polymer processing curve, but gives no information as the axes have not been defined. A single cutting trial can be used to identify the axes and render this graph useful as a processing curve, which can then forecast the cutting speeds available over a wide range of thickness and laser powers. Take the example where 3 mm thick nylon has been cut at 4.8 m/min at a laser power of 500 W. Assuming that a thickness of 3 mm is the middle of the range under evaluation, choose any point near the middle of the *X* axis and identify it as 3 mm. Now divide up the distance between this point and the (0,0) position of the graph equally to give 1,2,3 and then 4,5,6 mm etc., as shown in Fig. 4.8. The next step is to find the intersection of the curve with the thickness = 3 mm position. Having drawn a vertical line from the 3 mm position to intersect the curve, draw a horizontal line from the intersection as in Fig. 4.8 and identify this *Y* axis value with the correct cutting speed of 4.8 m/min. Once more divide up the distance between this point and the (0,0) position on the graph to give a well defined axis scale and then extend this to the end of the axis as in Fig. 4.8. The processing curve for nylon is now usable for the laser power of 500 W. (Compare the values it gives with Table 4.1.) If cutting speeds for a different laser power are

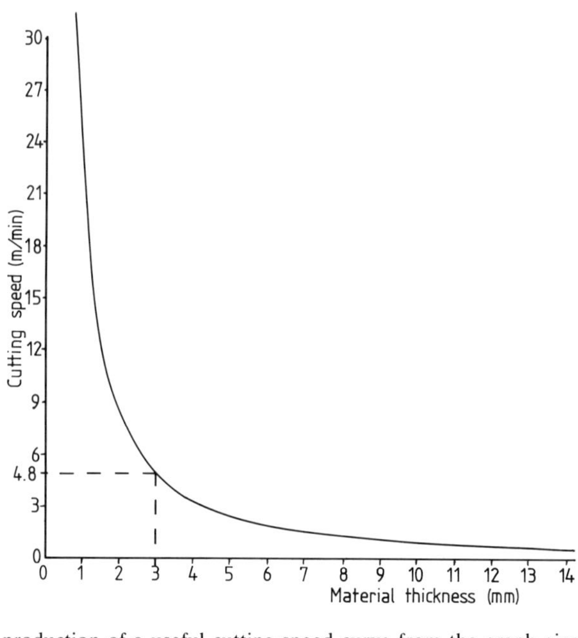

Fig. 4.8. The production of a useful cutting speed curve from the graph given in Fig. 4.7 and one piece of cutting data (see Sect. 4.2.5). Cutting data: 3 mm nylon cut at 4.8 m/min at a laser power of 500 W. This approach becomes more reliable if a number of thickness versus cutting speed results are used to produce an "average" graph.

required, simply change the Y axis (cutting speed) values proportionally with the increase or decease in power (e.g. double the cutting speed reading for double the power, etc.).

The interesting point about this technique is that it makes no difference where you place your original cutting result on the graph. If 3 mm is one of the thickest sections you are likely to cut then your information can be put towards the right hand end of the X (thickness) axis. If 3 mm is a relatively thin section for your envisaged application then your cutting data can be used towards the left hand side of the graph. The cutting results given by the graph will not be affected by your choice of position for the original information as the curve shows the relationship between thickness and cutting speed, which remains constant. It should be borne in mind that the method relies upon the accuracy of the original cutting trial and that the reliability of the information generated will improve if a number of tests are carried out. Each test can be used to generate a separate curve and if these are then overlapped (using the same axes scales) to produce a less sharply defined but more reliable graph.

One final point about the reliability of the method is that caution must be exercised if the forecast cutting result is at a thickness or power (or both) which differs from the actual cutting trials by a large margin. A result for 3 mm material at 500 W will give a good approximate forecast for 10 mm material at 1200 W or 1 mm material at 100 W for example, but cannot be relied upon for 0.1 mm material or for powers above about 2 kW where laser mode types become less predictable. It must also be borne in mind that each laser power will have a maximum thickness of material for good quality cutting, a feature which will not be shown up by this forecasting technique. It can generally be assumed that, in practice, a higher power laser will have a large focused spot diameter, a feature which will reduce expected cutting speeds and give a broader kerf width.

Development of the Graphical Method Using a Simple Formula. Approximate cutting speed forecasting can be carried out by using a minimum of cutting trial results in combination with the following simple formula:

$$V = PQt^{-B} \tag{4.1}$$

where

V = maximum cutting speed (m/min)
P = laser power (W)
t = material thickness (mm)
Q = an experimentally derived constant for the polymer using a particular laser–focusing optics combination.
B = an experimentally derived constant for the material.

If V is required in inches per minute and t is given in inches then the formula changes slightly to $V = PQt^{-B}/2$ (where $B = 1.35$).

Experimental work (Ref. 43, Chap. 11) has shown that the value of B can be taken as 1.35 over a wide range of polymers and so the equation is simplified to

$$V = PQt^{-1.35} \tag{4.2}$$

Values of V P and t can be obtained experimentally and Q can be calculated by inserting these in the equation.

This equation basically describes the following points: (1) the cutting speed is proportional to the laser power; (2) curves of cutting speed against material thickness have the same basic shape (given by B) for most polymers, although the values relating to the curves (given by Q) will be different.

The value of Q will be affected by the characteristics of the material and the focused beam. The important factors concerning the beam are the focused spot diameter and the energy distribution (mode) of the beam. Material will be cut at higher speeds if the size of the focused spot is decreased by, for example, reducing the focal length of the lens. For this reason Q cannot be established accurately for a material for all cutting machines, it is important that an experimentally determined value of Q is relied upon only for the particular laser–focusing lens combination used. Any established value of Q can, however, be used to give an approximation of the cutting speed, because cutting lasers are becoming progressively more alike in their output beam quality as inferior designs are discarded.

Using the Cutting Speed Forecasting Equation. Once a value of Q has been found by experiment it is possible to forecast the cutting speeds for a material over a wide range of thicknesses and laser powers. A single trial which establishes the cutting speed at one power at a single thickness is enough to give a value of Q, but of course, the reliability of the value and any subsequent forecasting will improve dramatically if a number of trials are carried out.

Example. Cutting polypropylene using a 500 W laser with a good Gaussian mode, an unfocused beam diameter of 12 mm and a 127 mm focal length ZnSe lens:

$$\text{Material thickness} = 9.525\,\text{mm}$$
$$\text{Maximum cutting speed} = 0.8\,\text{m/min (found by experiment)}$$

Putting these values into Eq. (4.2):

$$V = PQt^{-1.35}$$
$$0.8 = 500Q \times 9.525^{-1.35}$$
$$0.8 = 500\,Q \times 0.0477$$
$$0.8 = 23.85Q$$

Therefore $Q = 0.033$.

Putting this value for Q into the equation, the cutting speed at any laser power at any thickness of polypropylene can be estimated. For polypropylene using a 12 mm diameter Gaussian beam with a 127 mm focal length lens:

$$V = P\,0.033t^{-1.35}$$

Therefore at, say, 1200 W, with a thickness on 10 mm, the maximum cutting speed will be

$$V = 1200 \times 0.033 \times 10^{-1.35}$$
$$V = 1.77\,\text{m/min}$$

or at 200 W with a thickness of 1.5 mm,

$$V = 200 \times 0.033 \times 1.5^{-1.35}$$
$$V = 3.8\,\text{m/min}$$

4.3 Wood Based Products

It was mentioned at the beginning of this chapter that the cutting of plywood die boards for the packaging industry was one of the first commercial uses of CO_2 lasers. Since that time (1971) the CNC capabilities of the process have led to the development of its use in wood based industries as diverse as furniture manufacture and cigarette paper perforating. In the furniture industry lasers are commonly used to cut sections of up to 4 cm (1.5 in), although thicker sections are possible. Material savings are substantial and processing times are reduced to between 50% and 10% of those possible by traditional methods. Kerf widths are in the range from 0.2 to 1.0 mm (0.008–0.04 in) depending on the application and the cut edges are parallel and perpendicular. Lasers have found an application in the cigarette industry because the perforations produced by a pulsed CO_2 laser are flap and debris free. The finished product is stronger than its mechanically perforated counterpart as a result of the absence of edge tearing.

Cellulosic materials (wood, fibreboard, hardboard, paper, etc.) are cut by the same mechanism as thermoset plastics, i.e. chemical degradation (see Sect. 4.2.3). Temperatures in the cut zone can be anywhere in the range from 2000 °C to 3000 °C depending on the density of the material. Wood is composed of cellulose, water, lignin and trapped air in varying proportions, and the laser cutting process is affected by the relative proportions of these constituents. The cutting speed at a given laser power cannot be easily established for any particular type of wood because the water content varies from batch to batch and the cellulose density changes from area to area in a single sheet. The most profound local change in cellulose density is experienced when a knot in the wood lies on the cutting line. These knots, which are the cross sections of smaller branches radiating out from the original tree trunk, have a much higher cellulose density than the surrounding wood and can frustrate the full penetration of the cutting process. This can spoil the finished cut article, because any attempt at mechanical breakage of the partially cut knot will generally result in a fracture which will pass around the knot rather than through it. Plywood of any type is far more homogeneous than wood in its natural condition, and cutting speeds and quality can more easily be relied upon, providing the material is stored correctly. Wood stored in damp conditions absorbs atmospheric moisture which will reduce cutting speeds. Figure 4.9 shows the result of an experimental program in which more than 30 different woods were cut. All the wood samples had the same water content (12%) and thickness (10 mm). These samples were all cut at a laser power of 350 W, and a clear correlation between density and laser power is evident. In this case the density differences were due solely to variations of the cellulose and lignin content of

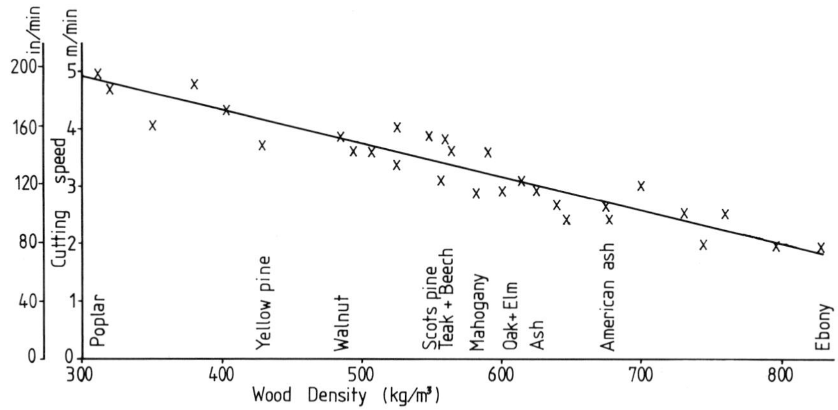

Fig. 4.9. Cutting speeds for a number of different wood types. Samples in all cases were 10 mm (0.4 in) thick and the moisture content was 12%. The laser power was 350 W. The relationship between density and cutting speed is obviously linear. (Cutting gas: air at 3–6 bar.)

the different woods. Increasing the density by increasing the water content has a similar effect of reducing the cutting speed.

Wood is most effectively cut using an air jet in conjunction with the laser. The oxidation of a proportion of the carbon content to carbon dioxide is an exothermic process which helps to supply the cut front with energy. The actual burning process is similar to that experienced in wood fires. Some workers have reported cutting speeds using an inert gas jet. This reduces cutting speeds and has little effect on the edge darkening (charring) process as the cellulose needs to be chemically degraded to carbon and, in any case, the wood itself contains a great deal of air.

The cut edge left by laser cutting is darkened by the presence of a residual carbon layer which is a result of the chemical degradation of cellulose. The level of darkening depends on the cellulose density of the original wood, for example teak, which is very dense, has a black cut edge, whereas a pine edge is only a few shades darker than the original wood. Darkening is minimised by using the highest cutting air jet pressures in order to strip the loose carbon dust off the cut edge. The quality of the cut edge is greatly superior to mechanical cut edges and is smooth, flat and splinter free as seen in Fig. 4.10. One common application for laser cutting is the production of decorative features for shopfitting etc. The speed and precision with which such complex patterns can be produced makes this type of work very profitable, compared with traditional carpentry methods. Laser profiling also allows engineers to look more closely at plywoods as a structural material for component manufacture. Plywood has a good strength to weight ratio and high rigidity, but has until recently been overlooked as an engineering material, mainly because it has been associated with inaccurate wood working methods. Now that working tolerances have been improved from millimetres to tens of micrometres with the advent of laser cutting, the material can be re-appraised.

Table 4.3 presents cutting speeds for various wood based products using a CO_2 laser in conjunction with a high pressure air jet. In the case of paper,

Fig. 4.10. The edge quality of laser cut wood. The edge is smooth, flat splinter free and generally darkened by the presence of a thin layer of carbon dust. This sample is 20 mm (0.75 in) thick pine.

Table 4.3. Cutting results for wood and wood-based products using a CO_2 laser

Material	Thickness mm	Thickness in	Cutting speed m/min	Cutting speed in/min	Laser power Watts
Poplar	10	0.4	5.0	200	500
Douglas fir	10	0.4	3.5	140	500
Yellow pine	10	0.4	3.2	128	500
Walnut	10	0.4	3.8	152	500
Cherry	10	0.4	4.3	172	500
Scotch pine	10	0.4	3.3	132	500
Beech	10	0.4	4.0	160	500
Teak	10	0.4	3.5	140	500
Mahogany	10	0.4	3.1	124	500
Oak	10	0.4	2.9	116	500
Ash	10	0.4	2.6	104	500
Ebony	10	0.4	1.2	48	500
Pine	6	0.25	8.0	320	1000
Pine	12	0.5	3.2	128	1000
Pine	20	0.8	1.6	64	1000
Plywood	6	0.25	7.0	280	1000
Plywood	12	0.5	3.0	120	1000
Plywood	20	0.8	1.5	60	1000
MDF*	6	0.25	9.0	360	1000
MDF	12	0.5	4.0	160	1000
MDF	20	0.8	2.0	80	1000
Hardboard	3	0.12	10.0	400	500
Hardboard	4	0.16	7.0	270	500
Corrugated card	3	0.12	25.0	1 000	500
Paper	0.1	0.004	$\geqslant$500	$\geqslant$20 000	500

* Medium density fibreboard.
Note: Generally use high pressure air as assist gas (3–10 bar, nozzle diameter 1–2 mm (0.04–0.08 in).

cutting speeds in the order of hundreds of metres per minute have been recorded at laser powers in excess of 250 W (Ref. 45, Chap. 11). Naturally the figures in Table 4.3 must be taken as rough guidelines only, as wood and wood based products are so variable.

4.4 Ceramics and Glasses

4.4.1 Ceramics

Ceramics are generally extremely hard, high melting point materials generated by the reaction of metals with elements such as oxygen, nitrogen and carbon. Commonly used ceramics in industry include alumina (Al_2O_3), tungsten carbide (WC), titanium nitride (TiN) and titanium carbide (TiC). Granules of carbide ceramics are often bonded together with additions of cobalt or a similar metal in order to make larger blocks.

Alumina is used as the substrate of micro-electronic devices as it is extremely stable and electrically non-conductive. Tungsten and titanium carbides are used in the production of cutting tools and drill tips where their extreme hardness is a great benefit. Other uses for ceramics include dies for high wear applications (wire drawing) and coatings for abrasive surfaces ("emery" cloth and non-slip flooring).

The ceramic of most interest to the laser cutting industry is alumina. This material is often produced in sheet form in a range of thicknesses (commonly 0.2–2 mm (0.008–0.08 in)) and components must be cut out of the sheet. As a result of the extreme hardness and brittleness of alumina sheet it is extremely difficult to cut by mechanical means. Although straight line cuts can be achieved using diamond impregnated slitting saws, it is impractical to cut complex profiles.

Laser cutting, being a thermal technique, is not frustrated by the hardness of the material, although the high melting point of Al_2O_3 (2015 °C) does result in low cutting speeds when full penetration cutting is required. In many cases, however, there is no necessity to cut the material fully, particularly where straight line cuts are required. The brittle nature of the sheet allows it to be easily broken along a chosen line if that line is first "scribed" by a pulsed laser to produce a row of overlapping indentations or blind holes. The inside surface of these holes is coated in rapidly resolidified cracked material, and so a continuous line of such indentations acts as a line of weakness which will easily break when stressed (see Fig. 4.11). Because the region of damage need not be very deep, the energy consumed by the process is small compared with full penetration cutting and process speeds can be high (15 m/min (600 in/min) or more).

The actual mechanism by which a pulse from a focused laser beam produces a partially penetrating hole in the surface of the alumina is still under investigation but one possible explanation is presented in Fig. 4.12 which shows that the dissociation of Al_2O_3 to aluminium and oxygen can suddenly increase the gas pressure on the laser generated melt and act as a driving force to expel the melt from the laser material interaction zone. The melt leaves the area as droplets and the walls of the resulting blind hole

Fig. 4.11. A magnified view of the overlapping blind holes produced by laser scribing. (Distance between hole centres is ~0.2 mm (~0.008 in).) The material can be easily broken along these scribed lines. (Photo courtesy of Coherent General.)

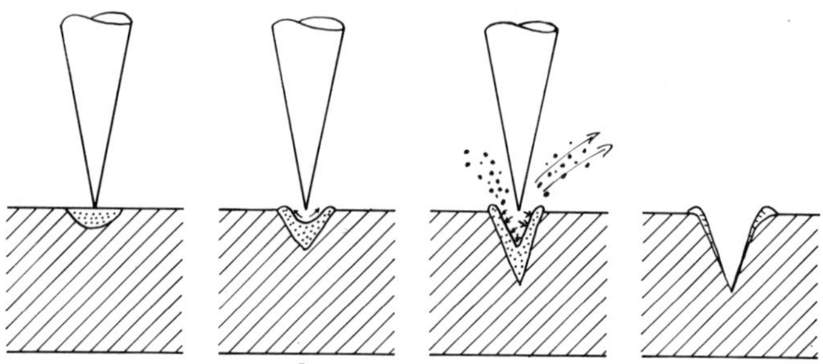

Fig. 4.12. A schematic of a possible mechanism for the scribing process in cross section. **a** The laser pulse begins and the material is rapidly heated and melted. **b** The severe thermal gradient across the surface of the molten pool (hot in the centre cool at the edge) gives rise to a surface tension gradient (low in the centre high at the edge) which draws the liquid away from the centre, creating a dimple. The shape of this central depression enhances beam absorption. The centre of the depression experiences a rapid increase in local gas pressure as a result of boiling and/or the dissociation of the oxide (e.g. Al_2O_3) to the metal and oxygen. **c** The rapid increase in gas pressure acts to push the liquid–gas interface towards the liquid solid interface, at the same time causing the ejection of the majority of the liquid. **d** The laser pulse ends, leaving a blind hole lined with rapidly solidified material.

solidify rapidly after the laser pulse ends leaving a microcracked highly stressed zone which can then be linked to another by the next laser pulse.

Full penetration cutting is required when complex or curved profiles need to be produced. The mechanism for material removal is a mixture of melt shearing and pressurised ejection. As in the case of hole drilling the sudden production of oxygen could increase the local gas pressure and act to force the rest of the melt out of the cut zone. Dissociation need not involve more than a few percent of the mass of material removed and does not affect the finished product. Dissociation is possible as a result of the high peak powers which the material is exposed to during a laser pulse. Continuous wave cutting can be carried out, but pulsed operation is generally preferred as the heat input to the area surrounding the cut is minimised and finer cuts achieved.

Laser drilling for the fabrication of hard dies for wire drawing, etc. is performed by repeatedly pulsing in one area until a full penetration hole is achieved.

Traditional ceramics, as used in the production of decorative tiles, can be cut by CO_2 laser by the melt shearing mechanism. The cut edge is covered in resolidified glass from the cutting process and is of good quality. The cutting process can, however, damage the glaze on the surface of the tile as a result of the severe thermal cycle and thermal stresses associated with solidification.

4.4.2 Glasses

Most glasses have a high absorptivity at the CO_2 laser wavelength of 10.6 microns. The limiting factor to the cutting of these materials is the possibility of micro- and macrocracking around the cut edge. This cracking can be caused by presence of a hot thermally expanded zone surrounded by the cold brittle substate or by solidification cracking. In the latter case the liquid glass is cooled too rapidly by conduction to the body of the workpiece. The stresses thus built up on the thin "weld" of material left on the cut edge exceed the fracture stress, and cracking occurs. If cracking of this sort is a problem, it can be alleviated by preheating the workpiece. This practice reduces the thermal gradients associated with the cutting process, and also renders the workpiece more ductile and less prone to fracture. Naturally the heat resistant glasses such as pyrex are less susceptible to cracking during cutting than other grades of glass. The most common industrial application of laser cutting glass is tube cutting. Glass tubing is cut into lengths which are subsequently formed into components such as light bulbs. The cutting process may be of the full penetration type, or scribing followed by mechanical fracture along the weakened line. If full penetration cutting is

Table 4.4. Cutting data for full-penetration profiling of ceramic materials using a CO_2 laser

Material	Thickness mm	Thickness in	Cutting speed m/min	Cutting speed in/min	Laser power Watts
Glass	1	0.04	1.5	60	500
	2	0.08	1.0	40	500
	3	0.12	0.5	20	500
Alumina	1	0.04	1.4	56	500
	2	0.08	0.6	24	500
	2	0.08	2.0	80	1000
Silica	1	0.04	0.6	24	1200
Ceramic tile	6.3	0.25	0.6	24	1200

required it is often achieved in a number of overlapping passes, as this reduces problems associated with thermal shock. For the same reason, pulsed rather than c.w. lasers are used. Research as shown (Ref. 3, Chap. 11) that it is possible to cut glass and ceramics by utilising the thermal shock generated by a lower power CO_2 laser to produce a controlled cracking reaction. This process has not been developed enough for industrial use, but does show some promise.

Table 4.4 gives an indication of cutting speeds available.

4.5 Composites and Miscellaneous Materials

4.5.1 Composite Materials

Many composite materials, for example glass filled epoxy resin, are difficult to cut by mechanical methods and are ideally suited to laser cutting. Cutting speeds and qualities are, in general, a compromise between the cutting characteristics of the materials joined to create the composite. For example a comparison between laser cutting nylon 6:6 and its glass reinforced counterpart yielded the results given in Table 4.5 at a laser power of 400 W.

The refractory nature of glass is obviously responsible for the higher temperature of cutting and the reduction in processing speed. At these elevated temperatures the nylon matrix is chemically degraded and cut edge charring is the result although it need not be deleterious to the performance of the cut product. The reduction in the number of stress raisers in the form of exposed fibres found after mechanical cutting could improve service life particularly where harder, epoxy resin matrices are employed.

Carbon reinforced materials have been laser cut with some success, but the edge quality depends on upon the carbon content and the degradation characteristics of the matrix. High carbon content materials are cut with difficulty and the high temperatures reached combined with the high thermal conductivity of carbon can result in heat affected zones of matrix degradation of unacceptable width. Problems of this sort can be minimised by using a CO_2 laser in its pulsed mode.

Composite materials cover such a large number of combinations and compositions that each example should be investigated on its own merits as to the feasibility of laser cutting. Table 4.2 presents a few examples to indicate the order of magnitude of cutting speeds, but it must be borne in mind that matrix/fibre proportions are rarely given in the literature. As these figures are unavailable and have strong influence on cutting speed, Table 4.2 must be regarded as only a guideline.

Table 4.5. The effect of glass fibre reinforcement on the cutting speed of nylon

Material	Thickness mm	in	Cutting speed m/min	in/min	Cut front temp. °C	Comments
Nylon 6:6	3.0	0.12	3.9	156	660	Typical thermoplastic cut
Glass filled nylon	3.0	0.12	2.6	104	1900	Slight edge charring

Laser power: 400 W.

4.5.2 Miscellaneous Materials

The range of materials which can be cut with a CO_2 laser is enormous. Published figures include cutting trials carried out on materials as diverse as boron epoxy composite and oatmeal biscuit! (It is not generally satisfactory to cut foodstuffs by laser as the process leaves a charred edge which has a distinctive, rather unpleasant taste.) Since the 10.6 μm radiation is readily absorbed by most non-metals including many naturally occurring substances (e.g. leather), localised heating and therefore cutting is generally possible.

Any manufacturer proposing to use laser cutting should consider three questions:

1. Is the cut quality satisfactory for my requirements?
2. Is the cutting speed high enough to match my production rates?
3. Is the process economically advantageous compared with present methods of production?

Many manufacturers are surprised to find that the answers to these questions are very often positive. As early as 1974, CO_2 lasers made an impact on the garment industry by their ability to cut cloth cleanly and quickly. Cutting speeds could be of the order of 100 m/min depending on the cloth weight. Edge fraying was eliminated and, for man made fabrics, a sealed edge was produced. At more moderate cutting speeds, several thicknesses of cloth could be cut at one time. Another customer with an unusual application for lasers required the cutting of carpet tiles to high accuracy for the production of inlaid mosaic type patterns and repeating designs over large areas. Mechanical cutting could not achieve features such as inlaid logos and also demonstrated cumulative errors in square size which disturbed simple designs if a large area was on display. Carbon dioxide lasers can also be used to strip various types of insulating material off copper cable without damaging the highly reflective copper itself.

It was stated earlier that isolated cutting results at various laser powers and materials thicknesses exist throughout the literature. Most metals and non-metals have been investigated at some point with varying, but generally high, levels of success. In view of the fact that the beam quality and power of lasers has been much improved over the past ten years, a potential new user of laser cutting should use published information only as a rough guide to cut qualities and speeds available at the moment. Projects which seemed unworkable only five years ago are now of great commercial interest (e.g. cutting mild steel at thicknesses between 14 and 18 mm). Bearing in mind the fluctuation in cutting performance between various grades of the same material, the only way to get exact information about the applicability of laser cutting to a new users production requirements is to have trial samples produced. Most laser manufacturers and a limited number of laser "jobbing" shops are capable of carrying out this service and the growth in the number of users over the past few years speaks for itself. Savings in waste material, processing time and post-cutting cleaning operations, combined with advantages such as the ability to have profiles cut after fabrication, have given tremendous financial returns to all types of manufacturers.

Chapter 5

Setting Up for Cutting

5.1 Introduction

The primary objective of setting up is to deliver the focused beam to the workpiece via an optical path of mirrors and lens(es) without distorting the purity of the beam by stray reflections off mirror mounts, etc. Once the beam is focused onto the workpiece in its purest possible state, it is then necessary to establish the focal position exactly and correctly locate this with respect to the material surface. The last step is to make sure the focused beam is aligned central to the nozzle and to set the cutting gas pressure and the cutting speed for the particular laser power–workpiece combination.

In summary, assuming the laser is tuned (see Sect. 5.6):

1. Steer the beam using mirrors to hit each mirror centrally and to pass through the centre of the lens.
2. Find the focal point and position it as required either on the material surface or a set distance into the depth of the material.
3. Arrange the nozzle to make sure it is aligned coaxially with the beam and is the correct distance from the workpiece during processing.
4. Set the laser power cutting gas pressure and process speed (in the case of a new process or material these parameters must first be established by trials).

The following sections will deal with these points individually and explain the techniques of setting up for cutting.

5.2 Beam Alignment

5.2.1 Introduction

Considering a rather difficult case for beam alignment, Fig. 5.1 shows a schematic for a cutting table with moving optics in the *X* and *Y* directions. To align such a system the laser beam is directed from place to place by tilting the mirrors which thereby steer the beam. This can of course be an extremely hazardous procedure if caution is not exercised, and must only be carried out by trained personnel. These notes should only serve as a reminder to such personnel and must not be considered as a full training

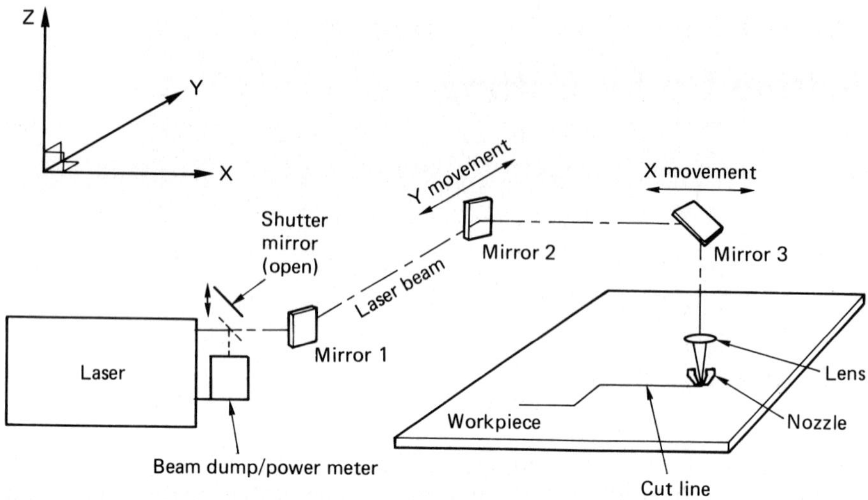

Fig. 5.1. A schematic of the optical chain of a cutting table with moving mirrors in the X and Y axes. The laser beam path is shown as a broken line. Mirror 1 is stationary, mirror 2 moves into and out of the page together with mirror 3 and its attached cutting head (lens + nozzle). Mirror 3 and the attached cutting head can move towards and away from mirror 2 (left and right in this diagram).

course. If you are in any doubt about a particular course of action, wait until a better qualified person can advise you. *Never* carry out a maintenance or setting up procedure alone, even if the actual work is to be done by a single worker.

The following notes explain beam alignment by two methods: (1) beam printing; (2) use of a secondary visible laser beam which can be projected along the same path as the CO_2 beam. These secondary lasers are fitted to many CO_2 laser cutting machines, particularly those with several beam steering mirrors. The low power red beam is generated by a lasing reaction employing helium and neon and the lasers themselves are often referred to as HeNes (pronounced heeny). If your equipment is fitted with a HeNe alignment device this reduces the number of beam prints needed, making the whole job of alignment quicker and safer.

5.2.2 Taking a Beam Print

Alignment of a CO_2 laser with its external optics requires that a number of "beam prints" are taken on sheets of transparent acrylic. The beam print is the evaporated impression left in the acrylic (perspex or plexiglass) after exposure to the beam. An example of such a print is given in Fig. 5.2. Two types of beam print can be used for different purposes:

1. A shallow print can be used to identify the position of the beam at any point along its path.
2. A deep print can be used to check the quality of the laser mode (see Sect. 1.4). These deep beam prints are often referred to as mode prints or mode burns (see Sect. 5.6.2).

Fig. 5.2. An example of a beam print made by firing the beam onto a sheet of acrylic for approximately half a second.

For the purposes of alignment the beam prints taken are the shallow, positional type, obtained by exposing the beam to the acrylic for approximately 1 s or less. The deeper beam prints will be discussed in depth in a later section dealing with tuning the laser.

The vapour given off during the production of a beam print has two unfortunate properties:

1. It is highly inflammable and can be ignited by the beam to give a yellow flame and a dense, sooty smoke which must be cleaned off any optical components.
2. Even in low concentrations it is highly absorptive of the laser light. If it is allowed to accumulate and become trapped in the flight tubes between mirrors during beam printing it will absorb a great deal of energy and diffract the beam, even though it is in low concentration and will not ignite. (Trapped vapour will also prevent successful cutting.)

To combat these two drawbacks it is best to use a vacuum cleaner to draw off the vapour from the area before it ignites. The vacuum cleaner tube can be hand held or clamped to one side of the beam print. For obvious reasons the tube should be of a substantial length if it is to be hand held: beam prints on the back of your hand are not only very painful, they are inaccurate! One alternative to drawing off the vapour through a vacuum cleaner is to clear it away with a compressed air supply. This method is only useful if shallow prints are being taken to plot the path of the beam, i.e. during alignment. If compressed air is blown onto a deeper beam print for mode observation, the results reflect the movement of the air over the melt rather than the laser mode and are therefore extremely misleading. When blowing air in the vicinity of laser optics care must be taken that the air is dust and oil free (most compressor air contains oil and water).

Exposure of the beam to the acrylic sheet can be controlled manually by operating the shutter control or, more accurately, by setting the CNC to either open the shutter for a set time (1 s or less should suffice), or to pulse the laser once or twice with the shutter open. The acrylic sheet should be clamped or taped in place, not hand held if at all possible. (The shutter is the first mirror the beam hits after it leaves the laser cavity. In its usual position it prevents the passage of the beam by deflecting it, usually into a power meter. When the shutter is activated it swings out of the path of the beam which then follows the optical chain to the workpiece.)

Operating the shutter for much less than half a second should be avoided as the movement of the shutter into and out of the beam path can give misleading results. Also shutters have been known to remain open if the open–close cycle is too short.

Briefly:

1. Clamp or tape a substantial sized (e.g. 20 cm × 20 cm × 5 mm) acrylic sheet in the path of the beam.
2. Position a vacuum cleaner hose on one side of the assumed position of the beam print to remove the vapour. Or, use a compressed air jet if the mode quality is not to be inferred from the resulting print.
3. For a positional beam print expose the sheet to the beam for approximately 1 s.

NB: The whole procedure is extremely hazardous; exercise a great deal of caution. Some of the covers which usually conceal the path of the beam will have to be removed to allow access for beam printing.

5.2.3 Alignment by the Beam Print Method

5.2.3.1 Aligning the First External Mirror

The term "first external mirror" refers to the first mirror the laser beam strikes after leaving the laser itself. (The shutter is considered to be part of the laser.) In many cases (e.g. a moving table *X*-*Y* system) this is the only external mirror and is used to direct the beam downwards from its original horizontal flight. This mirror in any system will generally be adjustable in two ways:

1. Movable as a whole in the *X*, *Y* and *Z* directions with a locking device when positioned.
2. Capable of controlled tilting to steer the beam.

It is the general position of the mirror with respect to the output beam which is the first thing to be set. The laser beam should hit this mirror as close to its centre as possible. To identify the positions of the beam on the mirror, beam prints can be used as follows:

1. Place a sheet of acrylic approximately halfway between the laser and the mirror and take a beam print. This will tell you if the mirror is mounted in approximately the correct position.
2. Fix a piece of acrylic close to the mirror mount and take a beam print. This acrylic sheet must be big enough to prevent accidental reflection of the beam off the mirror mount assembly and surrounding metal in the

Fig. 5.3. Cross wire targets which can be placed in the path of the beam to check alignment. The outside dimensions of such targets are designed to fit in selected positions in the optical chain, e.g. on mirror mounting blocks or instead of the lens holder.

case of bad misalignment. The beam print given will show whereabouts the laser would have hit the mirror (if at all).

3. Adjust the mirror position to centralise the beam on the mirror. (In extreme cases it might be necessary to adjust the laser's pointing direction by moving the whole machine.)
4. Check the centralisation by one of the following methods:
   (a) Take a beam print close to the mirror.
   (b) Remove the mirror from its mount and replace it with a disc of acrylic of the same size upon which a beam print can be made. (Don't forget to replace the mirror after this check is complete.)
   (c) With some systems it is possible to replace the mirror with a cross wire target (see Fig. 5.3). Beam prints taken a few centimetres beyond this target will show up any decentralisation very clearly (see Fig. 5.4). The resulting beam print has a superimposed image of the cross wire where it is has blocked the passage of the beam. It is important to have a target of this type which can be fitted in place of the lens (see next section). If cross wire targets are being used the beam print material can be card instead of acrylic (see Fig. 5.4). If card or heat sensitive paper are used the beam exposure time must be very short, or a single laser pulse can be employed to scorch the paper without igniting it.

5.2.3.2 Alignment of Subsequent Optics

Mirrors in the optical path after the first external mirror are not generally designed to be moved in the X, Y and Z directions as the beam can be centralised on them by tilting the mirror(s) closer to the laser. The procedure is basically the same as 1–4 above as far as the acrylic sheet manipulation and beam printing are concerned, but instead of moving the mirror to

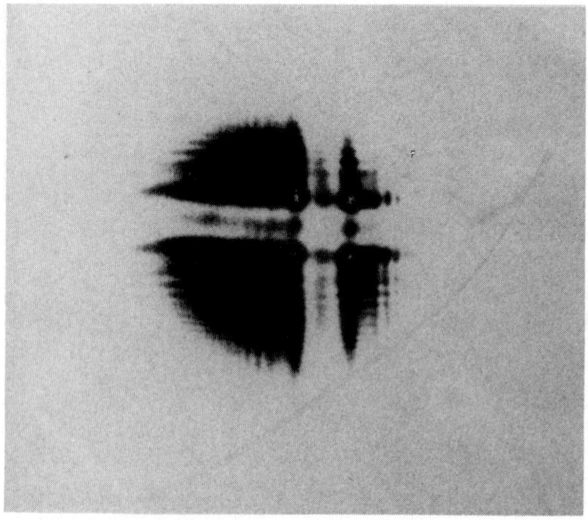

Fig. 5.4. An off axis beam identified by a beam print taken with a cross wire target in the optical chain. When cross wire targets are being used the beam print material can be either acrylic or card as shown here. When using card the beam exposure time should be minimised to avoid ignition of the material.

suit the beam position, the position of the beam on the mirror surface is manipulated by inclining the mirror next closest to the laser (i.e. for the second mirror the first mirror adjusts the beam position, for a sixth mirror the fifth is the adjuster etc.). Alignment must be carried out systematically one mirror at a time moving away from the laser.

For moving optics systems as shown in Fig. 5.1, alignment should initially be carried out with the mirrors close together with enough distance between them to allow positioning of targets etc. Having centralised the beam on mirror 1 it can now be centralised on mirror 2 by tilting mirror 1 (use beam prints and/or cross wire targets as before). Repeat the procedure by tilting mirror 2 to centralise the beam on mirror 3. Finally, centralise the beam on the lens holder.

Alignment of the beam with the lens is carried out without the lens in place. All transparent optical components such as lenses are highly susceptible to laser induced damage if they become contaminated in any way and so they must be kept well away from the potentially sooty environment associated with alignment. Mirrors can become similarly damaged if they become dirty, but they are less susceptible and can survive a couple of beam prints if they are unwittingly contaminated. All optics must of course be regularly checked for cleanliness, especially frequently during alignment operations. For comments on how to clean contaminated optics see Sects 6.2.8.3 and 6.2.10.

The inaccessability of the lens mounting in many cutting heads makes this an ideal opportunity for the use of a cross wire target (see Fig. 5.3) although simple beam printing methods can be used. A cross wire target may be manufactured by machining a sliding fit or threaded tube which will fit in place of the lens mounting. The cross wires are easily made by drilling four holes at 90° to each other in a ring around the tube and inserting steel or

Fig. 5.5. The type of beam print obtained if the beam is being inadvertently reflected off a mounting component of the optical chain. Mode prints of this type generally have a series of concentric semicircular ridges off to one side. (This beam print is made on card but acrylic would give a similar result.)

copper cross wires. The cross wires are only exposed to the defocused beam for a second or so at a time and therefore do not need to be made of any special material.

Once the target has been positioned in place of the lens assembly a sheet of acrylic or card can be placed in approximately the plane of the eventual workpiece (i.e. lying flat on the *X-Y* table). Beam prints combined with controlled tilting of the mirror nearest the lens can be used to centralise the beam. When the circle of the mode print is centrally intersected by the cross of the target the alignment is complete.

If the beam is very badly misaligned with the lens position the beam print may take on a number of strange appearances which are a result of the beam reflecting off the cross wire target tube or other metal parts of the cutting head. Figure 5.5 shows a typical example of this. Reflected, distorted beam prints of this type are typified by elongation in one axis and concentric semicircular ridges with the maximum energy density off to one side. This type of mode may give misleading indications as to the best direction for mirror tilt adjustment. In this case, tilt the adjusting mirror a substantial amount first in one direction, then the other, to relocate the unreflected circular mode. If necessary, take mode prints along the axis of beam propagation in order to establish the complete beam path.

The beam path is now centralised on the optics in the optical chain from the laser to the workpiece. This centralisation is, however, only true when the mirrors are in the positions they occupied during alignment. If the mirrors are now moved as far apart as possible it is unlikely that the beam will remain central on the lens position cross wire target. (If you do find that the beam remains in the centre, then the machine can be assumed to be fully aligned and you can move on to the next step – see Sect. 5.3.) The next step in the alignment procedure is to tilt the beam slightly until the path of the beam is exactly parallel to the mechanical movement of the mirrors.

Once this is achieved the beam will remain central on the lens and nozzle throughout the *X-Y* movement of the cutting head.

Fine tuning of the external mirrors is carried out by the following routine:

1. On a clean piece of card or acrylic make a mode print confirming the centralisation of the beam on the cross hair target with the mirrors in their initial alignment positions (i.e. close together with enough room between them to carry out beam print checks, etc.). Label this beam print.
2. Without changing the distance between mirrors 2 and 3 move them as far from mirror 1 as possible. Take another beam print through the cross wire target. (It is useful to make this print on the same piece of card or acrylic as the first. Remember to keep the orientation of the card, etc. constant with respect to the axes of the table for direct comparison.)
3. Label this print and make a note of the degree of eccentricity of the beam with respect to the cross (e.g. "beam centre is 4 mm too far west").
4. Use the tilt mechanism of mirror 1 to steer the beam until the centre of the beam print is off centre in the opposite direction by half the original amount, i.e. from 4 mm west to 2 mm east. (This is a hazardous procedure – use the smallest adjustment possible between beam prints. An apparently minor adjustment can move the beam completely off mirror 2 and a preliminary beam print should be made on a large piece of acrylic placed in front of mirror 2. Adjustments to the mirror should be made between beam prints, never with the beam on.)
5. Use the tilt mechanism of mirror 3 to centralise the beam on the cross wire target. Take and label a beam print to confirm this.
6. Return the mirrors to their initial alignment position and take a beam print.
7. If this beam print is central (±1 mm (0.04 in)) then the beam is aligned in this axis – move to step 9. If it is not central repeat steps 3–5 with the mirrors in their initial alignment position.
8. Repeat step 2. If the beam print is central then the beam is aligned in this axis – move to step 9. If it is not central, repeat steps 3–7 until alignment is correct in both positions within 1 mm (0.04 in). (Small amounts of misalignment can be accommodated by the self centring effect of the focusing lens.) The degree of misalignment should continuously decrease if the above routine is followed.
9. Once the first axis has been aligned a check should be made to identify where the beam falls on mirror 2. This is done by carrying out mode prints close to the mirror at both ends of its movements. The whole of the beam should impinge upon the mirror surface without touching any of the mirror mount assembly. Centralisation has to be sacrificed in favour of parallelism. If, however, it is not possible to make the beam parallel without unwanted clipping, it may be necessary to decentralise slightly the beam position on mirror 1. If this does not help, the system mechanics may be suspect.

10. Once the first axis has been checked, the second axis can be aligned. This can be carried out at any convenient position along the movement of mirror 2. First of all mirrors 2 and 3 must be separated as far as possible and a beam print taken through the lens position cross wire target. Alignment is then simply a matter of following steps 3–9 for mirrors 2 and 3 rather than 1 and 2. (Once the first axis has been set, mirror 1 should not be touched.)

5.2.3.3 Depolariser Mirrors and Collimators

Somewhere in the optical chain of most cutting machines there will be a mirror or pair of mirrors which act to "depolarise" the beam (see notes in Chap. 2 on energy input symmetry). This optical component may be called by its more correct name of circular polariser. Circular polarisation can be achieved by using one or two specially coated mirrors. If only one mirror is used it is probably the final mirror the beam hits before it passes through the cutting head. If a double mirror is used it may be earlier in the chain. The angle between the two mirrors in such a circular polariser is important and is often not adjustable.

These depolariser mirrors can be treated like any other mirror as far as the alignment procedure is concerned. Special care should be taken not to contaminate or damage them as they are generally expensive.

Some lasers have a collimator in the optical chain which helps to prevent the beam from changing its diameter by converging or diverging. Most collimators consist of a combination of lenses or a single flexible mirror whose focal length can be adjusted. Once again these optical components can be treated like the others for alignment purposes. Care must be taken to make the beam pass through the centre of each lens.

5.2.4 Alignment by Use of a HeNe Laser

5.2.4.1 General and Safety Comments

The use of a HeNe laser for alignment is extremely quick, convenient and safe, compared with the beam print method. The idea is to align a visible low power laser beam coaxially with the more dangerous, invisible CO_2 laser beam. The red HeNe laser beam can then be traced and manipulated along the correct optical path which will then be followed by the high power CO_2 beam.

Because their output lies within the visible spectrum, HeNe lasers are, naturally, a hazard if the beam is allowed to impinge on the eye. The HeNe lasers used for alignment are generally low powered and this means that they will not burn materials or skin, but great care must be exercised to avoid eye exposure by direct irradiation or by reflection off metallic objects.

Two important safety points to remember when using HeNe lasers are:

1. *Never* look directly down the beam.
2. Carbon dioxide laser safety goggles are generally no protection against the visible red HeNe laser output.

5.2.4.2 *Alignment*

To ensure alignment of the two laser beams the HeNe beam must be centred on two CO_2 mode burns, one within about a metre of the two lasers and one at a substantial distance from them. The more accurately this is done the greater will be the success of the whole procedure.

The HeNe laser is generally mounted on the much larger CO_2 machine near the shutter assembly, which releases the CO_2 laser light to the workpiece. The HeNe beam is brought into line with the main CO_2 beam by means of one or two adjustable mirrors which then reflect the HeNe beam off a prism (or mirror) attached to the outside of the CO_2 shutter. This prism (or mirror) then deflects the HeNe beam in the direction in which the CO_2 beam travels when the shutter is open. The two laser beams therefore travel the same path, but only one can be in operation at any one time, the deflecting prism or mirror being removed from the path of the HeNe beam when the shutter is opened.

In many cases the alignment of the two beams close to the shutter is factory set. However, the following procedure is for setting the alignment of the two beams in both the near and far positions.

1. Clear the area of all non-essential personnel (but do not work alone).
2. It is necessary to take beam prints within a metre or so of the laser output and at a more considerable distance. Therefore it will be necessary to remove some of the tubes or covers which usually contain the beam in its flight between the various optics. The distant beam print can be taken after reflection off several external mirrors if the beam is already roughly aligned with those. It is often more straightforward to remove one of the mirrors altogether and to let the beam escape from the usual optical path. In this way the minimum number of external optics are involved and complexities such as stray reflections are eliminated. Systems are often designed such that the first external mirror can be removed and the laser then fires in a straight line across the room which contains it. (This is of course a potentially hazardous operation which should only be carried out by trained personnel, taking as many safety precautions as possible.)
3. Fire the CO_2 beam to produce a beam print on a piece of acrylic placed within approximately one metre of the laser output. (Do not forget to use a vacuum cleaner to remove fumes.) This type of print will be referred to as the near target beam print.
4. Using the adjusting mechanism available (probably a prism or mirror combination) line up the red spot created by the HeNe laser on the centre of the CO_2 mode print. At this point the HeNe beam should be optically engaged whenever the CO_2 shutter is closed. (NB: HeNe lasers need at least 10 min to warm up – their pointing direction changes substantially over this time. Bearing in mind this and the warm up time for CO_2 lasers, any alignment should be carried out after both lasers have been warming up (lasing) for half an hour or more.)
5. Remove the acrylic sheet from the laser beam path and observe the path of the HeNe laser. This will be the approximate path of the CO_2 laser beam when the shutter is open. Depending on the arrangement of walls and access walkways in the area, either choose a convenient wall or build a

barrier of some sort at a distance of 5–10 m from the laser (a smaller distance may suffice, but accurate alignment will be more difficult).

6. Attach to this wall or barrier a substantial sized sheet of acrylic (e.g. 1 m × 1 m × 5 mm) and position this with the HeNe red spot at its centre. Using appropriate safety procedures, take a beam print on this sheet using the CO_2 beam for the shortest possible time (the distant target beam print). Do not try to use fume extraction or air jet cleaning of the beam print at this stage until the whereabouts of the beam are established. (Barriers should be constructed from non-reflecting material, brick or construction blocks are ideal, plywood will suffice.)

7. If a CO_2 beam print is successfully achieved in the expected area reinforce its image by super imposing a one second print and steer the HeNe beam by using the available adjustment until its red spot lies in the centre of the print. If no sign of a print is evident on the acrylic sheet place a similar size sheet much nearer the laser and trace the path of the CO_2 beam by beam printing. Repeat the beam printing procedure at increasing distance from the laser until the beam position at the far target distance is known and the target acrylic sheet can be repositioned.

8. Steering the HeNe beam to align it with the centre of the distant CO_2 beam print will have moved it slightly with respect to the near target beam print. Place a piece of acrylic at the near target position and take a beam print and compare the position of the HeNe red spot with respect to the centre of this print. If the move needed at the distant target to align the two beams was of the order of a few millimetres, the decentralisation of the two at the near target may be negligible. If the two beams are clearly non-concentric, however, further alignment is necessary.

9. Assuming that the beams are concentric at the far target but the HeNe is off centre at the near target it may be useful to draw yourself a diagram of the situation. Because the path of the HeNe beam is only of interest where it coincides with the high power CO_2 beam, we can take its point of origin as being its position of reflection off the final mirror or prism before it strikes the near target. The other adjustment mirrors/prisms (between the HeNe laser and this optic) make it possible to alter this origin position as well as the pointing direction of the beam. It will be necessary to alter both the origin position and the direction of the HeNe beam by manipulating these optics. Obviously if the HeNe is off centre by a couple of millimetres at the near target, but dead centre 6 m away, its pointing direction is almost correct. In this case move the adjustment optics not to tilt the beam, but to move the point of reflection off the final mirror and thus bodily move the beam sideways. The beam will now be central on the near target and, if the pointing direction has been untouched, a couple of millimetres off centre at the far target. A very slight adjustment of the tilting mechanism on the beam steering optics will now bring the beam to the central position on the far target once the near target has been taken out of the way.

10. Once again replace a piece of acrylic at the near target position, take a beam print and check for concentricity. If further adjustment is necessary repeat the procedure in 9 above.

When the beams are coaxial, the alignment of the HeNe beam with the external CO_2 optics will be equivalent to directly aligning the high power beam with them. All that is necessary is to steer the HeNe beam from the centre of the one mirror to the centre of the next, moving one mirror at a time away from the laser. Difficulties in seeing the red spot on the highly reflective surfaces of the mirrors can be overcome by laying a sheet of lens tissue over the mirror surface or by using acrylic sheet which has been roughened with emery cloth. As in the case of beam print alignment, the beam should be centralised on the first external mirror it strikes by adjusting the position of that mirror. In extreme cases of misalignment (e.g. when setting up a new machine) it may be necessary to tilt the laser machine as a whole (as the HeNe laser is bolted to the larger machine the two should remain aligned but this may have to be checked). Centralisation of the HeNe beam on subsequent mirrors is carried out not by adjusting the position of the mirrors, but by steering the beam by adjusting the tilt of the mirror next nearest the laser (see notes on beam printing alignment).

The HeNe alignment operation can be taken all the way down the optical chain from the laser to the workpiece including the focusing lens and nozzle. The difference in wavelength between the HeNe (0.633 μm) and the CO_2 beam (10.6 μm) means that the lens will focus the HeNe beam at a smaller distance from its own plane than the CO_2 beam, but this should not affect the coaxial alignment of the beams if they both pass through the centre of the lens. Nozzle alignment is, as we shall see in a later section, extremely critical and will probably need fine tuning to the CO_2 beam after alignment with the HeNe.

Certain machines employ a double lens arrangement in the optical chain to collimate the beam (i.e. to counteract its natural tendency to expand in cross section as it gets farther from the laser). Care must be taken to ensure that the HeNe alignment beam (and therefore its CO_2 partner) passes through the centre of each lens because the different focusing characteristics of the two beams could result in them becoming misaligned if they were off centre to the optical axis of this two lens collimator.

The HeNe beam which is emitted from the nozzle when the shutter is closed can be used to align a workpiece or jig with the eventual CO_2 beam. This can be especially useful for testing cutting programmes, etc.

5.3 Finding the Focus Position

5.3.1 Introduction

Although the situation should improve in the future, most manufacturers of lenses accept a 1 or 2% variation about the nominal focal length of any batch of lenses. This means that every lens must have its focal position identified to ensure optimum cutting results. When cutting steels or polymers an effective focal depth of approximately ±1% of the focal length exists, within which, cutting is fundamentally unchanged (i.e. material will be cut effectively over a range of ~2 mm at a distance of between 99 and 101 mm for a 100 mm focal length lens). This effective focal depth may be larger for various polymers, but highly reflective or difficult-to-cut materials may require more exact focusing.

A number of different techniques exist for establishing the focus position of a lens and two of the simplest and quickest will be described here: the blue flash test and the drilling test.

5.3.2 The Blue Flash Test

The blue flash test relies upon the fact that at the very high energy densities associated with a focused CO_2 laser it is possible to vaporise metal and, at the same time, to ionise that vapour. Ionised metal vapour (or plasma) emits high intensity ultraviolet and violet-blue light (blue flash). The ionisation of metal vapour involves energising it to the state where the metal atoms are stripped of one or more of their electrons and a cloud of positively charged ions and free electrons is produced. This requires a high energy input and will only take place at, or near, the focal point of a CO_2 laser beam where the energy densities are greatest. The basic technique involves irradiating a metal sample (commonly stainless steel) with short pulses whilst changing the lens–material distance until the focus position is established by one of two methods:

1. For low power machines or for machines capable of low power or very short pulses, the focus is that position at which the brightest blue flash and loudest noise is observed for an individual pulse.
2. For machines where such low power pulses are impossible or unreliable, a blue flash may be observed at a substantial distance from the focal point (e.g. 1 cm (0.4 in)). In this case the changes in the flash near the focal point will be too small to identify it accurately. The best answer to this problem is to defocus the beam in both directions (i.e. deliberately focus above and below the plane of the metal sheet) about the focus position until the blue flash is only just possible. The focus can then be assumed to lie halfway between these two positions.

5.3.2.1 The Technique

1. Set the laser to generate a single short pulse on demand. If this is impossible try computer controlling the shutter to open and close quickly at a minimum continuous power level. Shutters become unreliable below a certain cycle time so care should be taken to avoid being misled by differing laser "pulses". If short pulses by either method are impractical use the drilling method of focusing as it is not as heavily dependent on the generation of similar pulse energies.

2. After fitting the lens to the cutting head in the usual way put the head together but remove the cutting nozzle (to avoid the beam "clipping" it as it has not been aligned to the beam yet). Any nozzle–lens distance adjustment must be set as close to the lens as possible to allow the beam to be focused below the plane of the metal sheet without the head touching the sheet.

3. Use an air supply to the cutting head to protect the lens from spatter. Inert gas (Ar, He, CO_2, N_2) can be used if an air supply is not available – do not use oxygen. A high flow rate with low pressure can be expected because the nozzle has been removed. Excessive pressure will artificially suppress

vaporisation and might give misleading results. A general guideline is to have the exit pressure and flow of the air jet set at approximately the same level as the highest value you can achieve by blowing hard through pursed lips. (A large diameter short nozzle can be specially made for this test if the hole left by the removal of the cutting nozzle is too large.)

4. Position a sheet of stainless steel beneath the cutting head at a distance which is known to be beyond the focal point of the lens (e.g. for a supposedly 100 mm focal length lens make the lens–target distance 110 mm or more).

5. Expose the sheet to a single pulse from the laser and incrementally lower the cutting head (or raise the metal sheet) before pulsing again. Move the sheet laterally between shots or the second pulse will react artificially well as that area is already damaged. Continue lowering the head and firing the beam until the first blue flash is seen.

6. Vertical adjustment of the lens (cutting head)–material distance is usually made by a micrometer adjuster on the cutting head. When the lens focal position is near enough to the surface of the sheet to produce the first blue flash, make a note of the reading on the micrometer.

7. Continue lowering the lens and firing the beam on new areas of the stainless steel target until the blue flash event rises to a maximum and subsequently dies away – make a note of the micrometer reading of the last position at which a blue flash is possible.

8. Position the micrometer adjustment between the first and final blue flash positions. For most practical applications this can be considered to be the focal point. Checks can be made by repeating the test at a different pulse power to get a larger or smaller spread between the blue flash start and end points. One other method of checking is to use a different target material, for example aluminium should give a much narrower range than stainless steel. Do not use anodised aluminium as the aluminium oxide coating absorbs the beam too readily and obscures the result. Some materials give a white flash rather than blue but this does not effect the results.

5.3.3 The Drilling Test

This test works on the assumption that the smallest diameter beam (i.e. at the focus) will drill the smallest hole in thin sheet material. Set up as for the blue flash test with the nozzle removed from the cutting head and a substantial flow of air or inert gas. The target material should be either wood (or a wood based product) or acrylic sheet of a thickness between 1 and 0.5 mm. Wood or acrylic are used because they have a low thermal conductivity which reduces thermal "spreading" of the hole if a pulse is held for too long. Also, these materials are readily ablated and highly absorptive to the beam, and thus holes are created almost instantaneously.

Using a method of firing the beam into the sheet, then changing the lens–material distance and firing again into a new area of the sheet, the position of the focus can be established. It is important to pass from one out of focus range to the other in order to establish the middle point accurately. One method of accelerating this test is to mount the sheet of material at an angle to the horizontal. As the sheet is moved horizontally the lens–material distance changes automatically.

Once the focal position has been established it is useful to measure the distance from the target material surface to a point on the outside of the cutting head which cannot be moved with respect to the vertical position of the lens. In this way, adjustments of the nozzle–lens standoff and the overall position of the cutting head on its assembly will not cause confusion as to the whereabouts of the focal point.

5.4 Nozzles and Nozzle Alignment

5.4.1 General Comments

Whenever a CO_2 laser is used to cut a material, a nozzle is employed to direct a stream of gas coaxially with the focused laser beam in order to remove material from the cut zone. Nozzles vary in geometry, but the aim of their use is always the same; to direct a pressurised jet of a chosen gas into the cut. Bearing in mind the simplicity of this requirement it is perhaps surprising that scientific studies into the role of nozzles in laser cutting have revealed that the subject is very much more complex than it first appears. The reason for this complexity lies in the nature of the flow of a pressurised gas out of a small aperture. This subject will be dealt with in a separate section on gas dynamics after the more practical aspects of nozzle use have been reviewed. Most laser manufacturers supply nozzles which have the type of geometry shown in Fig. 5.6. These nozzles are simple to construct and are generally "designed" by trial and error methods rather than an in-depth analysis of the physics of gas flow. Very much more effective nozzles can be produced as a result of such analysis but they tend to be expensive and their performance deteriorates rapidly if they become contaminated with spatter, etc. Until a breakthrough is made in this subject the laser cutting industry will continue to be dominated by nozzles of the Fig. 5.6 type geometry, i.e. a hollow cone with a circular hole at the tip. These nozzles have certain peculiarities of behaviour which will be reviewed in the next section and explained in the gas dynamic section which follows it. For

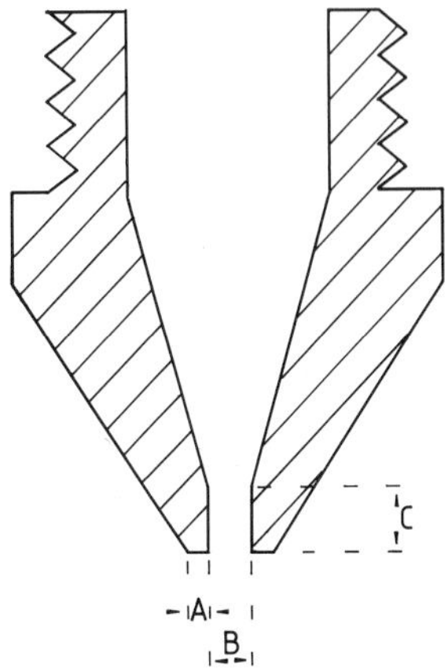

Fig. 5.6. A schematic cross section of a typical cutting nozzle. Nozzles of this types are usually made from copper. Approximate dimensions: A, ~1.0 mm (0.04 in); B, 0.8–2.0 mm (0.03–0.08 in); C, 1.0–2.0 mm (0.04–0.08 in).

example, increasing the supply pressure to the nozzle does not necessarily increase the pressure in the cut zone and can result in a sharp drop in effective cutting gas pressure. Even with these drawbacks the cutting industry has thrived using nozzles of this type.

5.4.2 Practical Considerations

5.4.2.1 Nozzle Design

Nozzles should be considered to be consumable items, although they may last for weeks or months even when cutting metals. Spatter from the process often adheres to the nozzle and repeated cleaning causes the tip to become worn. If polymers are the only materials being cut, spatter is not a problem and a nozzle can last indefinitely.

As spatter from metal cutting is a problem it is important that the nozzle is constructed of a material to which molten metal will not easily adhere. Copper fulfils this criterion very well and, being highly reflective to the laser radiation, is not readily damaged by the focused beam in the case of nozzle–beam misalignment. Brass is sometimes used as a replacement material for copper but is far more easily damaged by spatter and is not recommended, even though copper is much more difficult to machine.

As the nozzles are to be regarded as consumables and are to be made out of copper, it is appropriate to construct them as small screw on or slide in components. There is no need for the overall dimensions of the nozzle to exceed $1\,cm^3$ as only the outlet tip affects the gas flow into the cut zone.

Figure 5.6 shows the basic design for a typical cutting nozzle. As mentioned earlier, the gas flow from such nozzles is complex and varies according to the supply pressure. For this reason, trials need to be carried out to familiarise the operator with the performance of the nozzles. A range of outlet diameters will be required for different applications (e.g. 0.8, 1.0, 1.5, 2.0, or 0.03, 0.04, 0.06 and 0.08 in). Nozzles supplied by the laser manufacturer will have been developed by trial and error methods to give consistent results, but a change in design (e.g. by enlarging the exit diameter) may result in a nozzle which has superior or merely different characteristics.

5.4.2.2 Nozzle–Material Standoff

A major problem with nozzles of the Fig. 5.6 type is that the gas flow conditions just after the exit from the outlet can result in a sucking action rather than blowing effect. This is at first difficult to believe but can be explained by gas dynamics and is often clearly felt by approaching a (pressurised) nozzle with either a card or fingertip. At distances of more than a few millimetres there is a substantial force away from the nozzle. As the nozzle is approached to within 2 or 3 mm this repulsion is no longer steady and zones exist where it is obviously much weaker. In particularly bad examples there will be a zone very close to the nozzle (but not quite in contact with it) where there is a definite suction effect which can hold a piece of card, etc. It has been shown that the existence of a broad flat area (dimension "a" in Fig. 5.6) surrounding the nozzle outlet encourages this suction effect, but it can be noticed to some extent on all nozzles of this type. One way of alleviating this effect is to machine the flat tip into a shallow cone.

If the nozzle in use is known to have a region of 1 or 2 mm close to the outlet where severe changes in gas flow conditions exist, this area should be avoided by making the nozzle–material standoff larger than the depth of the fluctuating pressure zone. One other method to avoid such zones is to work with the nozzle just in contact with the workpiece, but this can lead to problems of nozzle wear. This subject of fluctuating pressure zones will be more fully discussed in the following gas dynamics section.

As with so many aspects of nozzle use, the correct conditions are most usually established by trial and error. Although the characteristics of the column of gas leaving the nozzle can be readily identified, the presence of the workpiece and the laser cut slot which forms the cut have effects on the gas flow which change the situation as far as local pressure fluctuations are concerned. A nozzle will therefore generally cut satisfactorily over a wide range of standoff distances. This is obvious when it is considered that much of the material cut commercially is not flat. A pitot tube of the type described in the gas dynamics section (Sect. 5.4.4) can be of great use when evaluating a nozzle and when considering an optimum standoff (see Fig. 5.9, Sect. 5.4.4).

The nozzle–material standoff is equivalent to the nozzle–focal point standoff in most cases, as the beam is usually focused on the material surface. It is important to realise that as the standoff is increased the likelihood of the converging beam clipping the edge of the nozzle is increased. Simple geometry can be employed to give a rough estimate of the maximum standoff allowed by diameter nozzle outlet. Nozzle–material standoffs for cutting work vary from 0 to 4 mm with 2 mm ± 0.5 mm satisfying most requirements. Many large laser–punch press combination machines use a nozzle material standoff of almost zero, controlled by automatic height sensing equipment.

5.4.2.3 Nozzle Gas Supply Pressure

Nozzle supply pressures for laser cutting range from 0.1 to 20 bar depending on the material being cut. Throughout this book, approximate figures for gas flow conditions are given for each material, alongside the processing data showing cutting speed against material thickness. One point to bear in mind, however, is that increasing the pressure to the nozzle does not necessarily increase the working pressure in the cut zone, and can even cause a sudden drop in working pressure. This effect is the result of the variation of pressure (or downwards thrust) of the gas jet with distance from the nozzle which was mentioned in the previous section. Increasing the supply pressure to the nozzle changes the positions of the low pressure areas and can therefore move, for example a "suction zone" into the plane of the workpiece (see Sect. 5.4.4).

5.4.2.4 *Choice of Nozzle Diameter*

As mentioned earlier, a range of nozzle diameters from less than 1 mm to 3 mm (0.04 to 0.12 in) is useful. Nozzles of about 1 mm diameter are used for steel cutting as they can supply the necessary pressure, whilst at the same time minimising cutting gas consumption. Slightly larger nozzles (~1.5 mm) are less sensitive to minor misalignment and changes in supply pressure. These can be useful when cutting steels at the thickest sections possible for a particular laser. The largest diameter nozzles can be used to supply low velocity cutting jets for acrylic sheet. In this case the liquid which solidifies to form the cut edge must be allowed to solidify without experiencing turbulence as a result of the gas jet (see notes on acrylic, Chap. 4).

5.4.3 Nozzle Alignment

There are several methods of nozzle alignment which give good results. The following section explains one method of coarse alignment which can then be used in conjunction with any of the three fine adjustment techniques reviewed subsequently.

In order to align the cutting gas jet coaxially with the focused beam there must be a method of adjusting the lateral position of either the nozzle or the focused spot. This adjustment will be one of two types:

1. Moving nozzle (see Fig. 5.7a)
2. Moving focused spot (moving lens, see Fig. 5.7b)

In case 1 the nozzle can be moved in the *X* or *Y* directions by adjusting screws or a combination of screws and springs. In case 2 the position of the

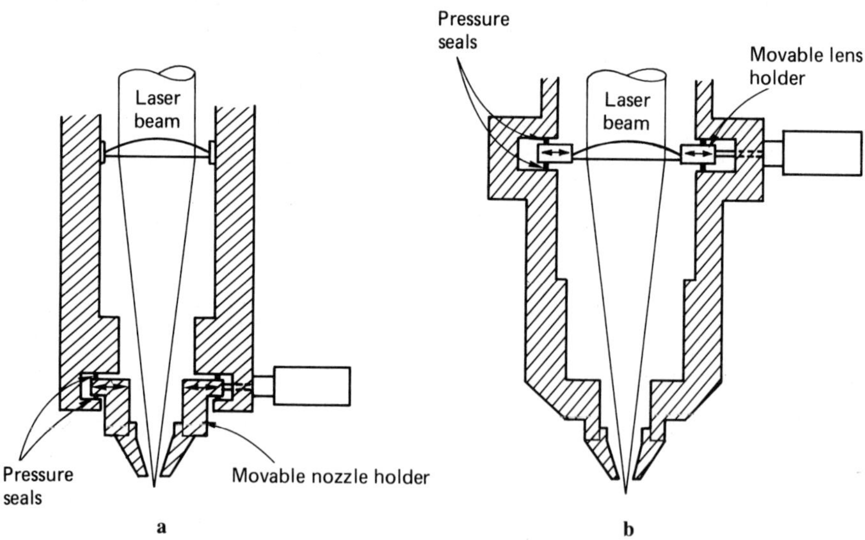

Fig. 5.7. The two basic mechanisms for nozzle alignment: **a** moving nozzle; **b** moving lens. This simple schematic shows movement in the *X* direction only. In practice movement would also be possible in the *Y* direction. (i.e. at 90° to the *X* axis, into the figure).

focal spot is moved by moving the lens and its mount. This latter method is often favoured by machine manufacturers as being easier to design and make, although it is technically inferior. It is true that as the lens is moved in the X or Y direction it will move the focal spot and therefore allow alignment to be carried out. However, moving the lens decentralises it from the path of the beam from the laser and undoes some of the earlier alignment work.

If this decentralisation is minor (e.g. ±10% of the unfocused beam diameter) then it will probably not affect cutting as long as the beam does not clip the side of the lens mount. On the other hand, larger levels of decentralisation will corrupt the laser mode at the focal point and may give poor cutting results. "Moving nozzle" cutting heads retain the purity of the mode by leaving the lens in its beam centralised position. The design of either movement device is not trivial as they both involve a pressure sealed sliding action.

5.4.3.1 Coarse Alignments

The aim of this technique is to use the laser to pierce holes in pieces of adhesive paper stuck across a large diameter nozzle. Misalignment is obvious from a comparison of the position of the hole with respect to the circular mark left by the nozzle on the adhesive paper. Adjustments can then be made to centralise the beam roughly. This procedure needs to be carried out only when the beam is too poorly aligned to pass through the exit hole of a cutting nozzle.

1. Remove the nozzle from the laser head and replace it with a specially prepared nozzle which has an exit diameter of 6 mm (1/4 in) or so. Alternatively the hole in which the nozzle is usually mounted can be used as the large diameter nozzle.
2. Set the laser so that it can be fired for a short time (0.5 s or less) at will (a "single pulse" option should be used if available).
3. Stick a piece of adhesive paper over the nozzle exit (use adhesive paper labels, or masking tape – not adhesive plastic).
4. When the beam pierces the paper a small amount of smoke will be produced which must not be allowed to come into contact with the lens. Any pressurised air supply will blow the adhesive paper off the nozzle before it can be laser pierced and it is therefore appropriate to attach a tube to the gas inlet to the laser cutting head and put the other end to your lips. Keep a very slight positive pressure on the chamber and then blow harder when the paper is pierced. This could also be done using a small hand held air pump of the type usually incorporated into lens brushes for cameras.
5. Fire the beam for a short time in order to pierce a hole in the paper, whilst at the same time blowing a small amount of air into the cutting head to remove the smoke.
6. Remove the adhesive paper and compare the position of the laser drilled hole with the circular mark made by the pressure of the nozzle

edge on the adhesive. Adjust the position of the nozzle or lens until the hole and the circle are coaxial by repeating the process (using fresh pieces of paper each time).

5.4.3.2 *Fine Adjustment*

Having approximately aligned the beam with the nozzle assembly it should now pass through a cutting nozzle without clipping the sides. After mounting a cutting nozzle the first test is to fire the beam for 1–2 s and then, when the beam is turned off, carefully touch the nozzle with a (wet) finger. If the nozzle is cold the beam is passing through cleanly. If it has become heated by the beam, some more coarse adjustment may be necessary.

Having established that the beam is clearing the nozzle without clipping the sides, one or all of the following techniques can be used to fine tune the alignment.

The Sparks Test. The aim of this test is to produce a small melt on the surface of a steel sheet whilst at the same time exposing this molten zone to a compressed air jet from the nozzle. The air jet will propel molten steel out of the area as sparks which will travel radially away from the melt across the top surface of the sheet. If the laser and air jet are coaxial the sparks should travel in all directions equally or in random directions. If, however, the laser is not central to the air jet, there will be a distinct preferred direction of travel for the sparks. The air in the jet moves radially away from its high pressure core and if this is to one side of the laser generated melt there will be a definite direction of flow across the surface of the melt as shown in Fig. 5.8. This subject is dealt with in greater depth in Sect. 2.1.7.4. If a preferred direction of travel is identified for the sparks, the nozzle should be moved in that same direction in order to align it with the laser beam or the beam should be moved in the opposite direction by moving the lens.

The technique:

1. Place a sheet of steel below the nozzle at the focal point of the lens or just below it.
2. Engage a pressurised air jet (~1 bar) through the nozzle and fire the beam for approximately half a second taking a note of the direction taken by the sparks. (Adjust the laser power to prevent the laser penetrating the sheet.)
3. Repeat 2 above three or four times on different areas of the sheet to ensure that your result is reliable.
4. Move the nozzle in the same direction as that taken by the sparks or move the beam in the opposite direction by moving the lens.
5. Repeat 2–4 until the sparks appear to have no preferred direction of travel. The beam and the gas jet can now be assumed to be coaxial, as the gas jet influences the molten pool equally in all directions.

The Grease Spot Test. This ingenious and simple test was developed by Fieret and Ward (Ref. 60, Chap. 11) and involves smearing the surface of a piece

Fig. 5.8. If the centre of the incident gas jet and the focal spot of the beam are misaligned, sparks from a laser generated melt will travel to one side (see also Fig. 2.17b). The sparks will travel along the axis of misalignment away from the core of the gas jet and therefore indicate the nozzle or focus spot repositioning necessary to achieve centralisation. (When the two are centralised sparks travel in all directions at random.)

of steel with a thin layer of black grease or any similar material. The grease is exposed to the air jet from the nozzle for a couple of seconds before the laser is fired for a single pulse or as short a time as possible. The resulting specimen reveals any misalignment by comparison of the position of the laser produced spot with the centre of the radial flow pattern formed in the grease.

The Beam Clipping Test. The aim of this test is to move the nozzle or beam in both axes to find the positions at which the beam just clips it. The central point between these positions for each axis should be the point at which the beam and nozzle are both aligned. The onset of clipping can be identified by drilling holes in a polymer (e.g. acrylic). The internal reflection of a portion of the beam off the inside of the nozzle will damage the top surface of the plastic on one side of the hole. A more accurate method is to drill holes in a heat sensitive paper which will reveal any small amount of clipping by the presence of a "shadow" to one side of the hole.

The technique:

1. Using a material which is sensitive to beam clipping (e.g. acrylic, paper or heat sensitive paper), repeatedly drill holes with the laser and an air jet whilst incrementally moving the nozzle or lens in one direction. Stop this movement when nozzle clipping is evident and make a note of the position of the nozzle/lens adjustment. (NB: Lasers with a power in excess of 1 kW may cause internal damage to a copper nozzle if clipping is allowed to continue for more than a fraction of a second.)

2. Reverse the direction of movement and repeat the drilling process until clipping is noticeable at the other extreme of this axis. Once more make a note of the adjustment position.
3. Move the nozzle or lens on this axis until it lies half way between the two clipping initiation points.
4. Repeat 1–3 for the other axis.
5. Repeat the whole operation as a check. The nozzle should now be equidistant from the four clipping initiation positions and can be considered to be centralised.

The final test for good alignment is the behaviour of the cutting process, particularly in the case of mild steel which is extremely sensitive to any misalignment of the beam with the nozzle. It is also very important to keep the nozzle free of any spatter, damage or debris which can upset the axial symmetry of the gas jet. Any degradation of this symmetry will have much the same effect as misalignment, the major symptoms of which are poor cutting in certain directions and burning in the case of mild steel. More details on the affects of misalignment are given in Sect. 2.1.7 and in the troubleshooting sections of Chap. 6.

5.4.4 The Gas Dynamics of Nozzles

5.4.4.1 Introduction

The following section is based almost entirely on the work on this subject published by B.A. Ward, J. Fieret and M.J. Terry of the UK Atomic Energy Authority Culham Laboratory (Abingdon, Oxon, UK).

A list of the relevant publications can be found in the bibliography section on gas dynamics in Chap. 11 (Refs 58–61). I would like to thank the Culham team for their permission to publish the following information in this slightly revised format.

A simple but elegant demonstration performed by Brook Ward at a conference in 1983 acts as a good introduction to the aerodynamic characteristics of simple conical nozzles with a circular outlet. The nozzle is set facing a pressure measurement device (representing the workpiece) as shown in Fig. 5.9. This pressure sensor can take the form of a small (~0.5 mm diameter) hole in a plate which is attached by tubing to a pressure gauge. If the gas supply to the nozzle is now engaged and the pressure slowly increased, the readings of the supply pressure gauge and pressure sensor gauge will initially rise in unison, although the supply pressure will always have a greater value. As the supply pressure is increased further, the pressure measured by the sensor gauge will continue to rise until a certain critical supply pressure is reached. A minor increase in the supply pressure now causes the pressure measured by the sensor to fall dramatically. This effect is demonstrated by Fig. 5.10.

As a second part of the demonstration the supply pressure is kept at a constant high value and the nozzle, mounted on the relevant mechanism, is moved towards the pressure sensor from an initial distance several millimetres away. Common sense would expect there to be a gradual increase in

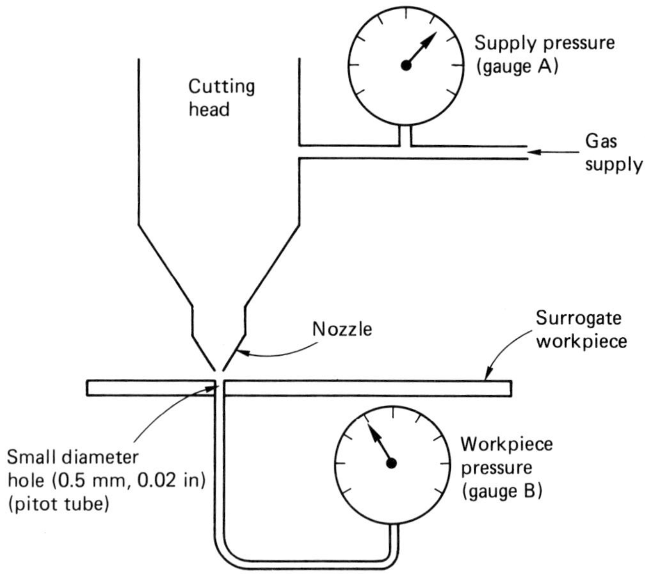

Fig. 5.9. A schematic of the experimental set up for monitoring the output pressure of a cutting nozzle for various supply pressures.

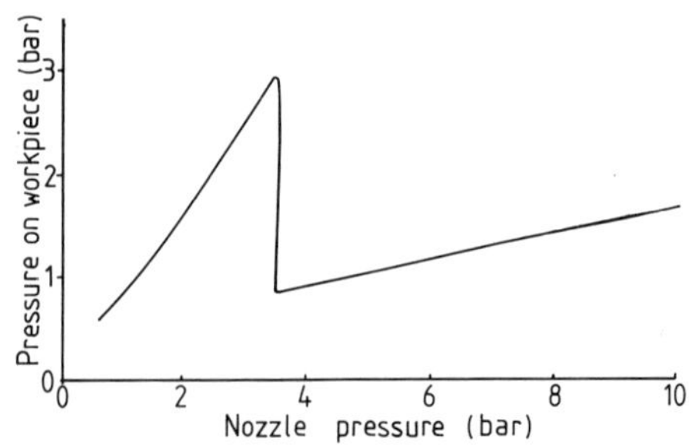

Fig. 5.10. The pressure on the workpiece exerted by a typical cutting nozzle as a function of supply pressure. Note the sudden decrease in "workpiece" pressure as the supply pressure is increased from 3.0 to 3.5 bar. These results are taken from the results of Ward and Fieret (Refs 58–61, Chap. 11). In this case the gas used was air through a 1.5 mm (0.06 in) diameter nozzle with a nozzle–workpiece distance of 3.5 mm (0.14 in).

"workpiece" pressure (as measured by the sensor) but this is not the case. As the nozzle becomes closer to the sensor the measured pressure is found to fluctuate considerably as shown in Fig. 5.11.

The reasons behind these rather unexpected results lie in the nature of the gas flow itself. Above a certain supply pressure the gas flow from a circular nozzle changes from subsonic to supersonic.

At low supply pressures subsonic flow persists, and in this case pressure gradients across the flow or along its axis are gradual and obey readily understood laws. At higher pressures the flow changes to supersonic, and

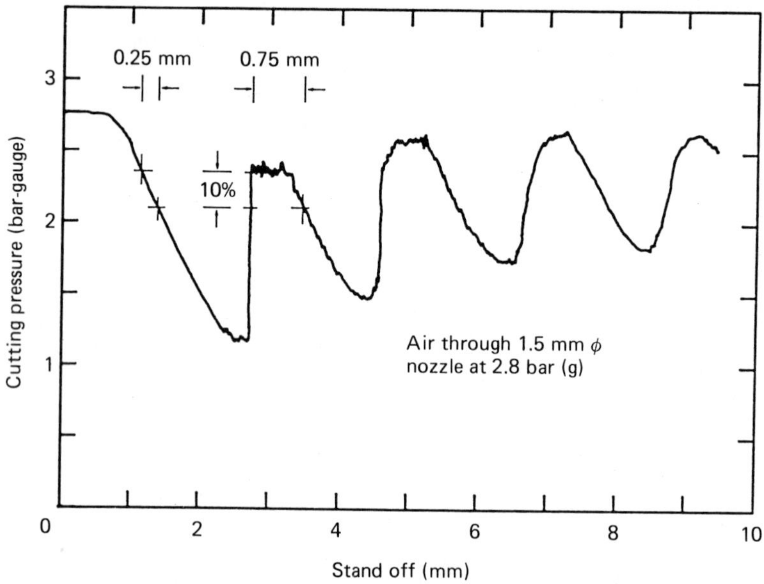

Fig. 5.11. The variation of the pressure on the workpiece with distance from a typical cutting nozzle at a constant supply pressure (Refs 58–61, Chap. 11). Once again the gas used was air through a 1.5 mm (0.06 in) diameter nozzle. The supply pressure was 2.8 bar.

sharp changes in local pressure are a dominant feature of the gas stream, either across its cross section or along its length.

5.4.4.2 The Practical Consequences of Supersonic Flow

Sharp variations in pressure downstream of a nozzle are a feature of most cutting gas jets. This being the case, it is very difficult to predict the pressure exerted at the workpiece for any supply pressure–nozzle variation. An increase in supply pressure can result in a sudden drop in local pressure at the workpiece surface. A change in nozzle diameter can also have unexpected effects as can a change in nozzle–material standoff. The work carried out by the Culham Group has concentrated on the dynamics of unrestricted jets and jets impinging on flat plates and therefore does not exactly reflect the flow conditions during cutting where the geometry of the cut slot will certainly affect flow. Nevertheless, practical work carried out by the Group does reflect a strong correlation between maximum cutting speed and the pressure incident on the plate surface for materials such as stainless steel and nimonic sheet. These materials form a high viscosity melt during laser cutting, and it is not therefore surprising that the physical removal of this liquid is a limiting factor to the cutting process. Figure 5.12 shows the results of cutting experiments carried out over a range of nozzle supply pressures. The sudden drop in workpiece pressure above a critical supply pressure (noted earlier – see Fig. 5.10) can be seen to be accompanied by a similar decay in cutting speed. Although this correlation between cutting

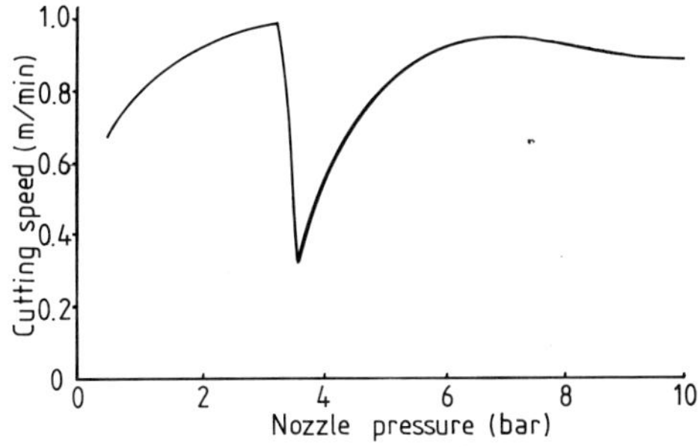

Fig. 5.12. The variation of maximum cutting speed with increasing supply pressure. The gas and nozzle characteristics were the same as for Fig. 5.10 and the close correlation is obvious. The laser power was 830 W and the material was 2.6 mm (0.1 in) thick PE16 alloy (Refs 58–61, Chap. 11).

speed and workpiece pressure may be expected when cutting by the melt shearing action of inert gas cutting, the situation is different when oxygen is used to cut steel. In this case the optimum pressure for cutting is not necessarily high. High pressures can lead to uncontrolled burning reactions in the cut zone. In any case, it is important that the gas supply to the cut zone is fairly constant in pressure and flow, and therefore conditions where rapid fluctuations in supply pressure are possible should be avoided. For example, if the nozzle–gas combination referred to in Figs 5.10 and 5.11 were being used as part of a repeatable industrial process it would be wise to use a below optimum but reliable supply pressure of 2.5 bar rather than the faster cutting but potentially unstable pressure of 3.2 bar.

It was mentioned earlier that nozzle standoff distances are usually set by trial and error methods. The use of a pitot tube can, however, replace this rather haphazard working practice. A pitot tube is merely a small diameter tube attached to a pressure gauge, as shown in Fig. 5.9. A simple way to produce one is to drill a small (0.3–0.5 mm) diameter hole in an acrylic sheet and then attach a plastic hose between the bottom of this hole and a pressure gauge (the hole can be laser drilled). This gauge should have approximately the same range as the gauge attached to the cutting head.

If the drilled hole is positioned under the nozzle in place of the workpiece, the nozzle standoff and supply pressure can be adjusted to give an optimum "workpiece" pressure which is stable over a reasonable range. In this way the low pressure zones in the supersonic stream can be avoided and nozzles can be assessed and compared. Central positioning of the pitot tube hole under the nozzle is very important. One aid to exact centring is to move the hole from side to side in both the X and Y axes until the central maximum pressure zone is identified.

5.4.4.3 *The Transition to Supersonic Flow*

The transition of flow from subsonic to supersonic is determined by only two factors: the ratio of supply pressure to the atmospheric pressure outside the

Table 5.1. Gas supply pressure for supersonic flow

Gas	Pressure (measured on gauge)		
	Bar	psi	kPa
O_2	0.89	12.9	89
Air	0.89	12.9	89
Ar	1.05	15.2	105
He	1.05	15.2	105
CO_2	<0.89	<12.9	<89

nozzle and the number of atoms in the gas molecule. The transition will occur when

$$\frac{P_g + 1}{P_a} = (1 + 1/n)^{1+n/2}$$

where
P_g = nozzle supply pressure (bar; as measured on gauge)
P_a = ambient pressure (bar)
n = number of degrees of freedom of the gas molecules

n is determined by the number of atoms which make up the gas molecules as follows:

He, Ar (monatomic) $n = 3$
Air, O_2 (diatomic) $n = 5$
CO_2 (polyatomic) $n > 5$

Assuming an ambient atmospheric pressure of ~1 bar (14.5 psi or 100 kPa), Table 5.1 shows the supply pressure above which each gas will flow in a supersonic manner. Note that this value is independent of nozzle diameter.

At pressures above these thresholds the flow rate of a gas through a nozzle can be easily calculated from

$$V = 8.2d^2(P_g + 1)\,\text{l/min}$$

where V is the flow rate (or gas consumption of the process), d is the nozzle diameter (mm) and P_g is the nozzle supply pressure (bar; measured on gauge).

At pressures just in excess of the threshold for supersonic flow a system of standing shockwaves will be established following the pattern shown in Fig. 5.13a. At higher pressures this intersecting shock wave becomes curved and a Mach shock disc is formed as shown in Fig. 5.13b. When a jet with these characteristics impinges on a workpiece the slipstream and radially diverging flow can form a stable vortex ring on the surface as in Fig. 5.14. The existence of such a vortex on the surface of a workpiece could interfere with the cutting process.

The work of Ward, Fieret and Terry has led to the investigation of the benefits of using nozzles with non-circular exits which has generated some interesting results. The reader is referred to the original literature for a more detailed description of the work (Refs 58–61, Chap. 11). One of the

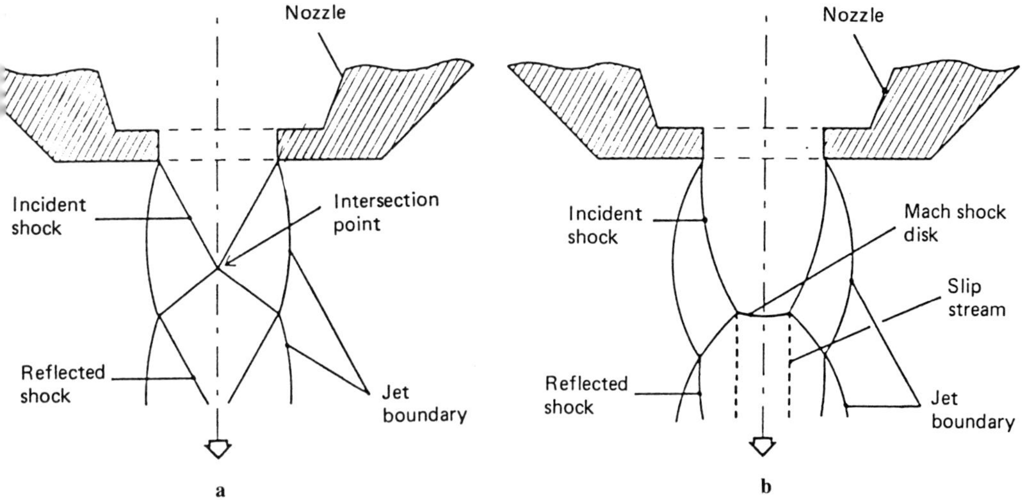

Fig. 5.13. Shock structures in a supersonic jet: **a** regular intersection; **b** at higher pressures the Formation of a Mach shock disc (Refs 58–61, Chap. 11).

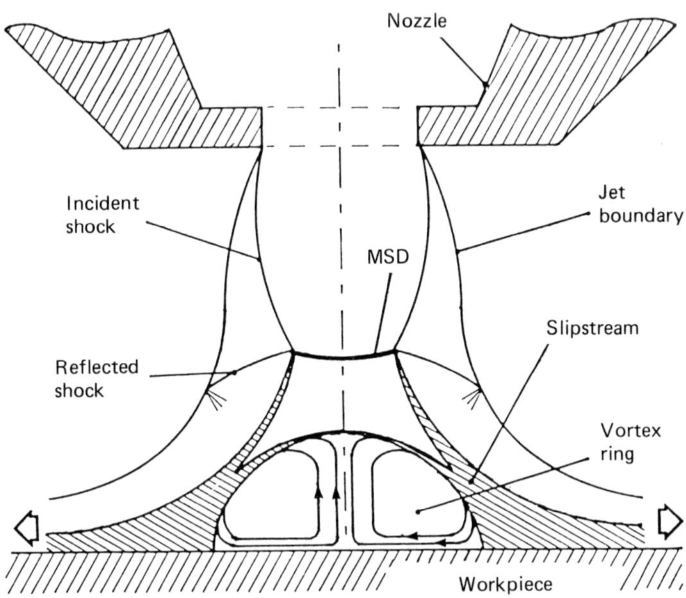

Fig. 5.14. The formation of a vortex ring on the workpiece surface. This type of flow could interfere with the cutting process (Refs 58–61, Chap. 11).

most promising of the new developments is the use of a coander nozzle, which is in fact a nozzle within a nozzle. The shock wave effects of the high pressure central nozzle are alleviated somewhat by the surrounding lower pressure air jet produced by a coaxial annular nozzle.

This subject of nozzle gas dynamics is still the subject of investigation and in future may well lead to the development of nozzles capable of accelerating and improving the laser cutting process.

5.5 Jigging

During laser cutting there is no mechanical contact between the cutting tool (the laser beam) and the workpiece. A great advantage of this situation is that jigs do not require the strength and rigidity necessary for other cutting methods. A new approach to jig design can therefore be developed making use of materials unusual to the engineering environment, such as wood, acrylic, adhesive tape and spring clips.

If components are to be cut out of large sheets, jigging will usually consist of clamping the sheet to the *X-Y* table (to avoid it sliding when the table executes sudden changes of direction) and supporting it on a matrix of metal spikes set into the base of the work table. These spikes (or nail bed) support the sheet under the cutting head but allow debris to leave the bottom of the cut zone freely. In most cases supports for laser cutting must be positioned away from the cutting line in order not to impede the flow of material out of the bottom of the cut zone.

If, on the other hand, the cut is to be carried out on prepared blanks, it is necessary to position the blanks carefully at the same time as supporting them. Taking the simple case of a rectangular blank, the best way to locate it is to cut a rectangle the same size out of a material of similar thickness which has been previously firmly clamped to the table. The cut rectangle's position is known and accurate location of the blanks in the rectangular hole is assured. Support can be afforded as before by a nail bed. An easier to use but less positive method of location is merely to cut an "L" shape out of a preclamped sheet and use the intersection of the *X* and *Y* axis cuts as a reference point. The blank to be cut is merely pushed into the "L" and can be either left unclamped or kept in position in a number of ways from clamping to adhesive taping.

5.5.1 Methods of Supporting the Workpiece

As mentioned earlier, it is generally important to support the workpiece without blocking the removal of material out of the bottom of the cutting zone. This support can be achieved by a number of methods, including the following.

Nail Beds. Nail beds consist of a matrix of spikes or rods which support the material to be cut. In many cases the spikes are individually removable in order to ensure a blockage free cutting path. The spikes are pointed at their top end to minimise the support – workpiece contact area. Spikes of this sort are often manufactured out of aluminium alloys to minimise damage if the laser comes into contact with them.

Honeycomb or Mesh. The use of a honeycomb or mesh to support the material is generally more applicable to cutting non-metals than metals. The cut line is, of course, intersected by the honeycomb or mesh at various points and, although the narrow nature of the intersection does not seriously interfere with material ejection from the cut, a build up of ejecta on the mesh is inevitable. In the case of polymers this build up is not serious as it will be easily removed by the laser on its next passage. Acrylic does not create a build up of material because it is ejected from the cut zone as a gas (see Chap. 4). If metals are cut, the honeycomb or mesh will become blocked eventually and will have to be replaced. The presence of the aluminium honeycomb will, naturally, be responsible for a slight reflection of the beam at the cut intersection points. This reflection can give rise to small perturbations of the cut quality on the bottom of a cut in, for example, acrylic sheet, although this need not necessarily affect the product.

5.5.2 Laser Cut Supports

In certain cases where cutting is to be carried out on prepared blanks, particularly when the blanks are formed into a three-dimensional geometry, it is preferable to abandon the use of a nail bed support in favour of a laser cut support specific to the job in hand. It is extremely useful that any jig which is laser cut out of preclamped material has its exact position and alignment to the movement of the machine automatically registered by the same CNC system which will carry out the eventual work on the blanks.

The simplest case of such an application is when the blank is too small to be supported by the spikes of the nail bed. A quick answer to this is first to cut the outline of the shape in a clamped sheet. This will identify the position in the *X-Y* plane, but in order to support it in the vertical plane it will be necessary to re-cut that outline slightly undersized and concentric with the first cut, in a sheet clamped or stuck (double sided adhesive tape) to the underside of the first sheet. The jig will then consist of an outline of the blank with a lip round the bottom edge on which the blank can rest. Blank insertion and removal from such a jig can take several seconds and so it is advised to produce a multi-component jig if continuous output is required. For complex geometries where more support is needed, one method is to build a jig which will hold the workpiece on a preclamped sheet and then cut holes in the sheet and jig to allow free removal of material from the bottom of the cut zone (e.g. if a 100 mm circle is to be cut from the component, cut a 120 mm circle from the jig at that position).

The workpiece can then be positioned on the supporting jig and cut as usual. Do not forget that jigs can be made of wood or fibreboard and nailed or glued together as long as they do not come into direct contact with the cutting process. In the most difficult cases it can be far simpler to separate the job into two operations with a jig for each. For example, if a number of small control panel fascias have been printed as an array on a perspex sheet and it is required to cut various switch holes from each fascia and then separate them; from a jigging point of view it might be convenient to carry out the "switch hole" cutting and then to separate the panels along one axis only as a first operation. Complete separation by cutting in the other axis

training has been provided by the laser supplier. Direct manual contact of such a block could result in fatal electrocution, if the mounting block is part of the high voltage circuitry.

The tuning operation itself is extremely simple and needs only care and patience:

1. Begin by choosing an adjustment axis and, after noting the initial output power and taking a mode print for future reference, tilt the mirror by the smallest amount possible in one direction. (See following section on mode prints.)
2. (a) If the power rises, continue in that direction until further movement causes a drop in power. Once the power has started to drop reverse the tilt to achieve the maximum output.
   (b) If the first adjustment made reduces the output power, reverse the direction of movement. If the power also falls in this direction the adjustment of this axis must have been optimised before tuning began so return to the original position.
   (c) If the first adjustment reduces the output power but reversing the direction of adjustment increases it, continue in this reverse direction until a maximum power is obtained. It is always necessary to continue adjustment until the power begins to drop again in order to identify that the peak value has been passed. Once the power has started to fall the optical adjustment must be reversed to achieve the maximum value.
3. Repeat operations 1 and 2 for the other tilting axis.
4. Repeat 1–3 to ensure that the optimum tuning conditions have been established. If this is the case the power should drop if the optic is tilted in any direction.

5.6.1.2 Tuning the Back Mirror

The back mirror is the mirror at the opposite end of the laser cavity from the output window. In some, but by no means all, cases this mirror is also user-adjustable for tuning. Adjustment is carried out exactly as for the output window (see previous notes) but certain points must be taken into account:

1. It is possible to misalign the beam completely by adjusting the two mirrors in an undisciplined way. Adjust only one mirror at a time as, using more than one mirror, it is possible to move the axis of the beam with respect to the central axis of the optical system. If the beam is bodily moved off axis in this way it may not be possible to re-optimise the mode without a complete internal re-alignment (see notes at the end of this section).
2. Adjustment of the back mirror can have a considerable effect on the pointing direction of the laser and may require the re-alignment of the external optics (see Sect. 5.2).

As a practical rule adjust one mirror at a time in one axis at a time to avoid confusion which could lead to a complete corruption of the mode; this would necessitate re-alignment.

Tuning should only be carried out when necessary (i.e. when the output power or mode are known to be in poor condition) and should involve the minimum number of mirrors, preferably only the output window or back mirror.

5.6.2 Mode Prints

In order to check the mode quality of the laser before and after tuning it will be necessary to take mode prints (or mode burns) in blocks of acrylic. Mode prints differ from beam prints (see Sect. 5.2) in that they are primarily used for mode analysis rather than the identification of the path taken by the beam. Mode prints in acrylic are often greater than 5 mm deep and it is therefore important to keep a supply of thick (≥15 mm) transparent acrylic sheet cut into convenient sizes. Mode prints should be taken at a position 1–5 m away from the laser output window with the minimum number of reflections off external optics. If possible, external mirrors should be removed from the optical path to allow an uninterrupted passage of the beam from the laser to the mode print position. In some cases one or two mirrors must be used to steer the beam to the print position and it must be borne in mind that any scratches, dirt or misalignment of these mirrors will corrupt the mode print. (Caution: disrupting the usual optical chain of mirrors and the removal of covers exposes the personnel involved to the danger of being burned by the (invisible) laser beam. Adequate safety precautions must be taken; see Chap. 7.)

Mode prints should always be taken at the same approximate position and should be labelled and kept for comparison and reference. The print itself is taken in much the same way as a beam print (see earlier in this chapter) except that the laser is fired for between 2 and 5 s to give a deep print. The depth of this print should not exceed twice its width if unrelated print shaping effects caused by the evaporation mechanism of the acrylic are to be avoided. A good choice of depth/width ratio is between 1:1 and 1:2 as in the example shown in Fig. 5.16. When taking the mode print never blow the vapour away with an air jet, always draw the vapour off using a vacuum cleaner (see Sect. 5.2). The flow of air from an air jet into the mode burn would change the geometry of the eventual print and give misleading results.

In summary, mode prints:

1. Are taken in much the same way as beam prints – see Sect. 5.2.2.
2. Should be taken at a distance of between 1 and 5 m from the laser output.
3. Should not be taken after reflection off external optics (if possible).
4. Should be taken at a set position and kept for comparison and reference.
5. Should have a depth to width ratio of approximately 1:1.
6. Should never have the vapour removed by blown air, suction should be used.

NB: Taking a mode print is an intrinsically dangerous operation and training from the manufacturer should be sought. All possible safety precautions should be taken (see Chap. 7.)

Fig. 5.16. An example of a mode print in an acrylic block.

5.6.3 Use of a Laser Beam Analyser

Although mode prints in acrylic are a convenient and cheap method of mode analysis they are, by their nature, only an approximate measurement technique. Over the past decade a number of devices have been developed to measure accurately the energy cross section of the beam. These beam analysis methods generally rely on sampling a small proportion of the beam and measuring this low powered signal by electronic methods. One of the most ingenious and easy to use devices of this type is the Laser Beam Analyser (LBA) developed by W.M. Steen and G.C. Lim (Refs 64 and 65, Chap. 11).

The beam analyser samples the beam by rapidly passing a molybdenum wire through it several times per second. The wire reflects a small sample of the laser light onto a pair of photodetectors during its passage through the beam. The geometrical arrangement of the wire and the two detectors is such that the energy density of the beam across two perpendicular cross sections is measured. These two energy cross sections give an accurate sample of the mode of the beam which can be displayed on a dual beam oscilloscope. The Laser Beam Analyser uses only 0.1% of the beam for sampling and does not affect the mode of the beam. It can therefore be used for on-line, in-process measurement of the quality of the mode.

5.6.4 Laser Alignment

Laser alignment is a major optical operation and should only be carried out by fully trained personnel using the correct equipment. Alignment is often necessary after transportation and installation of the laser to a new location.

For the best modern machines even this operation is unnecessary and alignment is carried out and fixed at the supplier's factory.

The only other times when a laser may need re-alignment is if the internal optics are replaced or if the tuning operation goes seriously wrong and the alignment is accidentally disrupted.

The aim of the alignment is obviously to position all the mirrors in the laser cavity correctly. The techniques used involve removing all the mirrors and then replacing them one at a time, checking their alignment by using a HeNe laser or a specially designed telescope. One method uses a HeNe laser pointing into the mirrorless laser cavity from the output window (now removed) end. Using targets with holes in their centre which sit where the missing optics would be, this beam is aligned centrally to the line of discharge tube(s) leading directly to the output window position.

The mirror next in the system to the output window is the first one to be replaced in the laser and is used to steer the, now reflected, HeNe beam down the next length of laser cavity (see Fig. 5.17). The beam is aligned centrally by steering it through the central hole of a target positioned where the next optic along would normally be. Once this alignment is achieved, this second optic is replaced and used to steer the HeNe down the next length of the system. The method is repeated mirror by mirror until the back mirror (the one at the opposite end of the laser from the output window) is replaced. Once the back mirror is replaced the output window is the only optic missing from the laser. The back mirror is aligned by using it to steer the HeNe beam directly back along the path it has taken to reach it. This part of the operation is rather difficult as it means looking for the weak red spot generated by the reflected beam in an environment dominated by the bright incoming beam.

If the target is removed from the output window position and a piece of white card with a hole in its centre (for passage of the beam) is placed in front of the HeNe laser, the weak reflected spot which should be apparent on the card can be steered until it coincides with the output of the laser.

The situation is now that the beam from the HeNe laser passes down the central axis of the optical cavity and is reflected back on itself along that line. All the optics are now aligned except for the output window which can

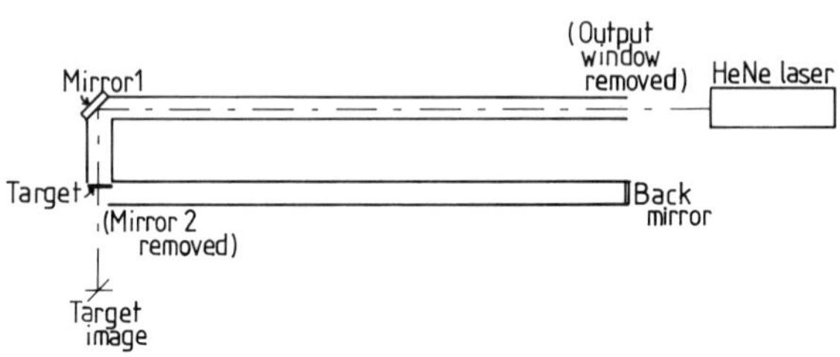

Fig. 5.17. Internal laser alignment: with the output window and mirror 1 removed a HeNe laser is shone centrally down the first length of the laser cavity. (This concentricity is checked by the use of targets with holes in the centre.) The second operation is to replace mirror 1, remove mirror 2 (and replace it with a target) and centralise the path of the reflected HeNe beam with the second length of laser cavity and so on. The eventual aim is to have the HeNe beam travelling centrally down the entire laser cavity and reflecting off the back mirror to travel back along the same path.

be replaced in the system. The output window will reflect a good proportion of the HeNe beam back upon itself and can be approximately aligned by steering the reflected spot to the output of the HeNe laser. Although this means that the outer face of the output window is aligned perpendicular to the direction of travel of the HeNe beam, this does not necessarily mean that the inner reflective face is aligned with the internal cavity, as these semitransparent optics are sometimes deliberately machined to avoid their faces being parallel to each other. Parallel faces in such an optic can give rise to deleterious optical effects.

Now the laser has been aligned, it can be fine tuned by switching on and tuning for maximum power and best mode as usual. Engaging the vacuum system for the first time may, however, disturb the position of one or more mirrors.

Chapter 6

Troubleshooting

6.1 Introduction and Checklist

Although the reliability of laser cutting machines has risen dramatically over the last decade, problems associated with the cutting process will inevitably arise from time to time. Most of the problems are related to incorrect parameter settings (power etc.) or the accumulation of dirt on optical components and the nozzle. This section hopes to guide the way to the most effective method of tracking down a problem and then eliminating it.

In any laser cutting system there are a number of factors which can give similar symptoms of poor cutting. For example, an incorrect power setting can give similar bad results to those obtained from an incorrect cutting speed or a damaged focusing lens. It is important therefore that the check list followed when the machine is cutting badly begins with the "easy to check and correct" variables and then progresses logically towards the more time consuming activities if the earlier checks prove satisfactory. It is extremely frustrating to find you have spent half an hour checking and fine tuning the laser mode when all you needed to do was clean the nozzle.

The checklist shown in Table 6.1 is a reasonable starting point for any laser system, although some users may find it necessary to add other tests as experience of a particular machine reveals specific areas of weakness. The checklist is accompanied by notes on each check. The timescales indicated are for rough guidance only and refer to an experienced operator on an "easy to access" machine.

It is assumed for the purposes of this checklist that the operator has found the laser cutter incapable of doing a standard type of job with which he is familiar (e.g. cutting 3 mm mild steel at 2 m/min). The checks should be carried out in the order they are presented in the list to prevent any time being wasted.

6.2 Notes on Checklist

6.2.1 Nozzle Contamination

The most frequent cause of poor cutting performance is adherent material on the nozzle tip which tends to deflect the cutting gas flow to one side and thus upset the coaxial symmetry of the laser–cutting gas combination. A

Table 6.1. Troubleshooting checklist

	Check and correct	Time involved (minutes)
1.	Nozzle contamination	1–2
2.	Laser power level and type (c.w. or pulsed)	1–5
3.	Cutting speed	1–2
4.	Cutting gas type, pressure and/or flow rate	1–2
5.	Nozzle material standoff	1–2
6.	Nozzle type, condition and alignment	1–10
7.	Material specification	1–5
8.	Lens type, condition and alignment	10–20
9.	Laser mode quality	20–40
10.	External mirror condition and alignment	5–60 per mirror

quick demonstration of this deflection mechanism can be carried out if the stream of water flowing from a tap is lightly touched on one side. The stream is clearly deflected towards the "touched" side. In the same way any artificial extension of the nozzle on one side by a particle of spatter will drag the gas flow over to that side and disrupt cutting in the same manner as a misaligned nozzle. The front surface of a nozzle can generally be easily cleaned by gentle scraping without having to remove it from the laser head. Most nozzles are made of copper to which most types of spatter (iron oxides etc., or polymers) will not adhere well.

6.2.2 Laser Power Level and Type

Checking that you are working at the correct power level and correcting this, if necessary, only takes a few minutes if nothing is wrong with the laser. The job takes on a new dimension if the laser is not generating the required power at a familiar setting. If, for example, the laser usually generates 1 kW at a particular setting but is only producing 800 W then the most likely causes are:

1. The machine has only recently been switched on and has not had time to warm up. In this case wait a few minutes – many lasers need half an hour or so to warm up, particularly after a cold night or weekend of inactivity. If production is urgent, cut at a reduced speed until the power rises.
2. The helium supply is running low. Towards the end of a helium or mixed gas bottle, but before the pressure is low enough to engage the helium supply interlock, it is common for the power to fall slightly. Change the helium or mixed gas bottle.
3. The laser needs tuning. Vibration and natural thermal cycles have acted to change the alignment of the internal optics of the laser (see Sect. 5.6).
4. The laser needs servicing. At this point trained personnel are needed as a large number of phenomena can be responsible for power loss. The most common reason for power loss is the build up of contaminants on the surface of the internal mirrors, which then need to be cleaned. If a

particular laser is susceptible to this problem it may be most effective to have a member of the user's staff trained in this aspect of maintenance by the laser supplier.

6.2.3 Cutting Speed

Check the feed rate in the computer program and the percentage feed rate setting. After establishing that the correct speed is being employed in the program, observe the process to estimate the actual velocities – many CNC machines default to a maximum feed rate if the information is not correctly entered. Most CNC machines have a maximum speed at which they can execute curves which decreases with decreasing radius of curvature. In the case of bolt holes (e.g. radius 4–1 mm) this maximum speed may be less than the programmed feed rate, which can therefore result in burning of the area around the hole. An excessive feed rate will result in incomplete penetration of the cut and will produce a combination of sporadic burning and surface melting. Cutting too slowly will result in burning of the workpiece, particularly on corners and small details.

6.2.4 Cutting Gas Type, Pressure and/or Flow Rate

Most laser cutting machines can be set up to cut with oxygen, air, an inert gas or a mixture of these. If jobs are changed frequently it is possible to attempt cutting using the wrong gas or gas mixture. Cutting quality is also dependent on the rate of supply of the gas to the cutting zone. The cutting head will be supplied with a pressure gauge or flow meter which should be preset at the required level. If the gas pressure to the nozzle is low the supply bottle should be checked, pressures from such bottles fall steadily in the last 5 min before they become empty. Another consideration is leakage: sparks from the cutting process or laser reflections can perforate the plastic hose generally used to carry gas to the laser head.

It is useful to have both a pressure gauge and a flow meter as together they give far more information than either one can in isolation. For example, if the pressure rises slightly but the flow drops, then the nozzle is partially blocked. When cutting mild steel with oxygen, a great deal of care must be taken to optimise the gas flow into the cut zone. Too high a gas pressure will result in burning, particularly on corners and details. On the other hand, too low an oxygen supply to the cut zone will not effectively remove the molten material and the result will be incomplete cuts which may also experience sporadic burning. The burning in this case is the result of the failure to eject the hot melt out of the cut zone. This melt subsequently heats the surrounding material by conduction and thus the workpiece becomes overheated and burns. When cutting polymers, or other non-metals, with air, the gas flow should be great enough to propel the inflammable vapours produced out of the cut zone before they ignite. Too high a gas flow may, however, tend to produce rougher cut edges as a result of increased turbulence in the cut zone. When cutting acrylic with a polished edge the gas must be set at a minimum level to prevent edge "frosting" (see Sect. 4.2.2).

6.2.5 Nozzle–Material Standoff

Nozzle–material standoff needs to be checked for two reasons:

1. As the distance between the nozzle output face and the focal point of the beam is known, the nozzle–material standoff is a good quick check of the position of the focus with respect to the workpiece surface.
2. The nozzle–material standoff to some extent determines the gas flow into the cut zone at a certain supply pressure to a particular nozzle (see Sect. 5.4). If a nozzle is replaced by another which is not identical to it the standoff may change. On most machines it is possible to alter the nozzle–lens distance to re-optimise the process. The optimum nozzle–material standoff may be different, depending on the nozzle diameter used and the material to be cut. In most cases, however, the standoff will be from 0.25 to 3.0 mm (0.01 to 0.12 in).

Steels are particularly sensitive to the nozzle material standoff distance. Mild steel suffers from burning on corners or details if the nozzle standoff is not optimised, particularly if it is too large. On the other hand, stainless steel cutting becomes unstable if the standoff is too small. If the nozzle is too near the surface of the stainless steel the cut zone begins to sputter, and sparks can be seen leaving the top of the cut zone. The resulting cut edge is rougher than usual, and details will suffer from burning. The process can be stabilised by raising the nozzle. Whenever the nozzle–workpiece standoff is increased the diameter of the incident gas jet is effectively increased. This may result in a reduction in gas flow through the cut zone, and supply pressures may have to be increased to re-optimise the process.

6.2.6 Nozzle Type, Condition and Alignment

This is the first check in the list to require a dismantling operation, albeit a minor one. As continued production is the major requirement of this check list, wasted time should be kept to a minimum. If a supply of identical nozzles exists they should be kept close at hand, and a new one should be fitted to the laser whilst the old one is being examined. If cutting improves with the replacement of the nozzle, the old one can be cleaned and returned to good working order or, if the damage is extensive, drilled out to become useful as a large diameter nozzle or discarded altogether. Damage to nozzles ranges from minor ailments such as spatter inside the nozzle outlet, to mechanical damage caused by collision with the workpiece etc. With high power or pulsed operation machines the nozzle can be eroded or drilled by the laser itself if the nozzle is not correctly set coaxially with the beam in the first place. A nozzle outlet can be cleaned by manually spinning it on a loosely fitting allen key or twist drill – this is best done by pushing the allen key through the nozzle which can then be rolled across the palm of your hand. Another method is to thread a metre or so of string or cord through the nozzle. Holding the string taut the nozzle can then be pulled along the string simultaneously rotating it to prevent preferential wearing of any one side.

6.2.6.1 Nozzle Misalignment

Misalignment of the nozzle with the laser is a common cause of poor performance. The cutting process is extremely susceptible to small (0.2 mm (0.008 in)) imperfections in the coaxial nature of the nozzle–laser beam combination, particularly when mild steel is cut with oxygen.

A major symptom of nozzle misalignment when cutting mild steels is a secondary shower of sparks which travels away from the cut zone across the top surface of the steel sheet when cutting in certain directions. This effect is clearly shown in Fig. 6.1, and the poor cut quality associated with it is demonstrated in Fig. 6.2. To cure this effect, simply adjust the position of the nozzle by moving it in the direction the sparks are travelling (an experienced operator may be able to do this during cutting). If the nozzle is moved too far, an alternative secondary shower of sparks will be generated, travelling in the opposite direction. If required, the initiation positions of these two opposite showers can be noted and the nozzle eventually positioned half way between them. The reasons behind the secondary shower generation and its necessary adjustment are described in Sects 2.1.7.4 and 5.4.3. (On some machines the nozzle position cannot be adjusted. In this case the lens should be moved, in the *opposite* direction to that of the sparks, which will thereby move the position of the focal spot.)

Fig. 6.1. A secondary shower of sparks travelling across the sheet surface when cutting mild steel with oxygen. This effect is caused by the misalignment of the nozzle with the focused beam. The beam and gas jet can be recentred on each other either by moving the nozzle in the same direction as the shower of sparks, or moving the lens (and thus the focused spot) in the opposite direction. Cutting machines will have one of these adjustments available to the operator.

Fig. 6.2. The poor quality cut obtained when the nozzle and beam are misaligned. NB: In the case of misalignment of this sort the burning affects one side of the cut only. This could also be caused by a poor quality mode (see Sect. 2.1.7). If burning occurs on both sides of the cut the problem is more likely to be incorrect gas pressure or cutting speed.

Another major symptom of misalignment is a reduction in the major shower of sparks leaving the bottom of the cut zone when cutting in certain directions. In this case the problem is caused by the oxygen jet having its centre ahead of the focused beam. The cut quality is impaired by the dross adhesion when this condition exists, but the situation is easily rectified by identifying the cutting direction in which the shower of sparks is poorest and then moving the nozzle in the opposite direction (i.e. if the machine cuts badly left to right, move the nozzle slightly to the left or the lens to the right).

When cutting polymers or non-metals such as wood based products, axial misalignment of the beam and the gas (usually air) jet can have one of two effects depending on the degree of misalignment:

1. For minor levels of misalignment the type of lateral flow across the top of the cut seen in Fig. 6.1 can deposit material on one side of the cut. This effect can be seen in Fig. 6.3. In some cases this accumulation of condensed vapour or dust does not affect the finished component and can easily be wiped off. In other cases (for example uncovered acrylic) the hot vapour can damage the top surface of the sheet near the cut edge.
2. If there is a considerable amount of misalignment it is possible for a portion of the beam to reflect off the inside of the nozzle. This effect is often referred to as "clipping". The disturbance of the focal spot associated with clipping will result in lower cutting speeds and poor cut quality. One common symptom of clipping when cutting non-metals is a "shadow" to one side of the cut where the small amount of deflected energy causes surface damage. This "shadow" is demonstrated in Fig. 6.4.

Fig. 6.3. Deposition of molten material or vapour on one side of a cut as a result of nozzle–beam misalignment. As in the case of the spark shower the nozzle should be adjusted towards the deposit side of the cut or the lens moved in the opposite direction.

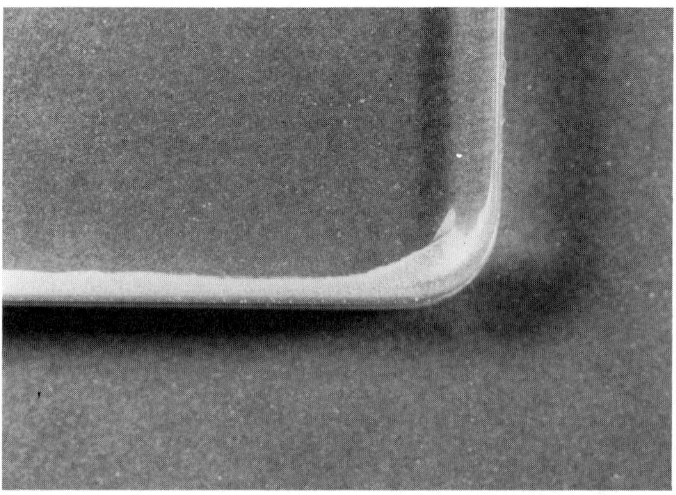

Fig. 6.4. Surface damage to a polymer sheet as a result of the beam clipping (reflecting off) the inside of the nozzle. In this case the stray beam has produced a frosted effect on the surface of a transparent acrylic sheet. The problem is only noticeable when cutting in certain directions (perpendicular to the axis of misalignment of the beam and nozzle).

6.2.7 Material Specification

If checks 1 to 6 have not solved the cutting problem, it is now appropriate to check the material specification before moving on to the more complex and time consuming checks.

The cutting process is very sensitive to workpiece chemistry and thickness, and problems can arise if either of these are changed. Confusion in the storage area of the work place is the most likely source of mistakes, although material suppliers are far from infallible. Material thickness is a critical variable of laser cutting, and although metals are generally very close to the required thickness, it is useful to note that this is not the case for many polymers. Possibly the worst offender is acrylic sheet which can vary by 10%–20% across a single sheet! Polymers and other non-metals can also have a wide range of chemical compositions under one name, e.g. "rubber" or "plywood". Cutting speeds can therefore vary by a factor of two or more (see Chap. 4).

One important point in the context of material specification is the difference between hot rolled (black) mild steel and cold rolled (bright) mild steel. At moderate laser powers (~500 W) the black oxide coating on hot rolled steel can slightly disrupt the cutting process to give an inferior quality cut (see Sect. 2.1.6).

6.2.8 Lens Type, Condition and Alignment

6.2.8.1 Lens Type

If a stock of different focal length lenses exists, it is possible for confusion to occur and a lens of incorrect focal length to be used to replace a damaged optic. When in doubt, a quick approximate check can be made by focusing the image of an overhead light source (striplighting, skylights etc.) on a workbench and measuring the lens–workbench distance. This test is only useful for ZnSe optics which are transparent at visible wavelengths, but it must be borne in mind that the lens will focus the infrared laser light to a different (more distant) focal position (e.g. a 127 mm focal length lens will focus daylight at a distance of ~100 mm). Laser optics which are opaque to the eye (e.g. GaAs) must have their focal position established by the usual laser methods (Sect. 5.3).

6.2.8.2 Lens Condition

Lenses can be made from a variety of materials which are transparent at the laser output wavelength of 10.6 μm. With the exception of the cheapest KCl (potassium chloride) optics these lenses are coated on both sides with anti-reflection (AR) coatings to eliminate reflection losses. The most common lenses used for CO_2 laser cutting are made from ZnSe (zinc selenide) with AR coatings which, when new, have a flawless surface and are transparent amber in appearance.

If a particle of dust or spatter attaches itself to the lens surface it will become rapidly heated by the beam and will damage the AR coating and the lens material in that area. In the worst cases the local thermal gradients set up around particles of dirt will crack the lens, causing laser light to be diffracted in many directions. More usually the dirt causes only surface damage, but its presence upsets the focusing ability of the lens. Each opaque area of dirt acts as a diffracting centre which changes the path of the beam around it as well as locally preventing passage of the light. The effect of dirt particles is not, therefore, simply to block out a small proportion of the light, but also to disturb the integrity of the beam in the focus position. This disturbance will have the effect of enlarging the focal spot size, thereby reducing the power density in the cut zone. The effect on cutting will be to reduce the cutting speed, enlarge the kerf width, increase the level of dross adhesion and, usually, to affect certain cutting directions much more than others.

Removal of the dirt (see next section) will generally reduce these problems to an acceptable level, but the damaged surface of the lens will continue to have a minor effect on process efficiency. If possible, a lens which has surface damage of any sort which is visible to the naked eye should be replaced.

6.2.8.3 Lens Cleaning

Cleaning of lenses is a delicate operation and can cause more damage than it repairs, as it is difficult to remove grit from the lens surface without scratching it. Scratches have the effect of a diffraction grating and lend to superimpose a line on the focused spot. Although it is still possible to cut "insensitive" materials such as stainless steels with such a beam, mild steel would respond badly and polymers would suffer from surface marking.

To clean the lens, first remove it in its holder from the laser cutting head and visually inspect it. If the dirt is loose on the lens surface, remove it by blowing. If this is successful, replace the lens assembly in the cutting head and consider yourself very lucky. If, as is more likely, the dirt is adherent to the lens surface, the next step to take is to wash the lens and its holder (the lens assembly) in a liquid. (There is generally no need to take the lens out of its holder to clean it and leaving the lens assembly intact reduces the risk of breakage.) The liquid used to wash the lens can be any of the following:

1. Alcohols – isopropyl alcohol, ethanol, methanol
2. Degreasing fluids – acetone, trichloroethylene
3. Water

The liquid used must be clean and grit free. Most of the fluids mentioned above will leave a deposit on the lens when they evaporate, and this is to be avoided. It is therefore recommended that a final rinse should be carried out in isopropyl alcohol which leaves no stain. The second best cleaning liquid in this respect is acetone. In the absence of these liquids the lens can be adequately washed in water and the final clean should be carried out by breathing gently on the lens and then wiping off the condensed water with a lens tissue. If the proper cleaning fluids are available, however, do not wipe the lens with anything until it is clear that the wash–final rinse method will not completely clean the lens.

If wiping is necessary, use lens tissues (available from any camera shop) or the cotton wool tipped sticks used more usually for baby hygiene (these are often lightly oiled and need to be degreased). Use plenty of cleaning fluid (preferably keep the lens assembly immersed) and try to free the dirt from the lens without subsequently dragging it across the surface. Only use each lens tissue or cotton wool tip for one wipe as they will accumulate grit and a second wipe will scratch the lens surface. As a last resort, use a (clean) fingernail to prise the dirt off the lens surface.

After removing the dirt by washing or wiping, rinse with isopropyl alcohol or acetone and allow time for it to evaporate *completely* (anything from 1 to

5 min) before replacing the lens assembly in the cutting head. If acetone or alcohol vapour becomes trapped in the flight path of the beam it will absorb and diffract the beam. This high absorptivity of the vapours is due to their hydrocarbon nature: most hydrocarbons absorb 10.6 µm radiation very effectively.

One final test to check for surface anomalies (including drying marks or smears from wiping) is to breathe on the lens and to watch the condensation pattern of your breath. If the condensation pattern is uniform then the lens is clean. Smears, etc., will show up as the surface disturbance they represent acts as a condensation initiator. At this stage, when the lens is grit free, a clean lens tissue can be used to wipe off the condensed water. Keep breathing and wiping until your breath condenses uniformly on the lens surface.

After the lens has been reinstalled the nozzle will probably need some realignment before cutting can commence. Whenever any part of the optical chain from the laser to the workpiece is dismantled, the components between the dismantled section and the workpiece will probably need some realignment.

6.2.8.4 Lens Alignment

If the beam is found to be off centre with respect to the lens, realignment will be necessary following the procedure laid down in Sect. 5.2. The misalignment can be identified by use of a cross wire target as described in the same section.

If the problem of misalignment is recurrent it will be necessary to identify the cause. In most cases this will be found to be due either to the movement of optics as a result of vibration, or thermal expansion/contraction of the cutting machine as a whole.

6.2.9 Laser Mode Quality

The quality and speed of cutting is highly dependent upon the mode of the laser as explained in Chaps 1 and 2.

A corruption of the laser mode is usually associated with a drop in output power, but this is not always the case. A mode print (see Sect. 5.6.2) will rapidly establish the approximate quality of the mode. As far as cutting is concerned, the most important aspect of the mode is that it must be as axi-symmetrical as possible in order to cut equally well in all directions.

Mode prints from earlier checks should be compared with the present results. If these prove to be similar and the overall output is the same at known preset values then the mode can be assumed to be satisfactory. If, however, the mode print is markedly different then the laser must be tuned according to the guidelines laid out in Sect. 5.6. If the laser is tuned it will probably be necessary to realign the external optics and the nozzle, as adjustment of the laser optics usually redirects the beam slightly.

If the mode and output power do not improve with tuning, then the problem is probably inherent in the laser itself. One fairly common problem is referred to as "thermal lensing". This phenomenon is most often experienced in machines of powers in excess of 1 kW. Thermal lensing involves the distortion of the output window as a result of its inner or outer surface becoming contaminated. If a layer of contaminant (oil vapour etc.) becomes attached to a surface of the output window it will begin to absorb the beam, and the subsequent heating will cause it to swell slightly towards the centre. This swelling changes the window shape to a lens and this will disrupt the lasing action inside the laser and the quality of the output beam. A quick check for thermal lensing is to take a few beam prints along the path of the beam every half metre or so. (See notes on taking beam prints in Chap. 5 and safety considerations in Chap. 7.) The change in diameter of the prints will give an indication of the rate of expansion or contraction of the beam. Most cutting lasers have a low divergence which means that the beam retains much the same diameter over several metres. If, however, you find that the beam reduces in diameter rather rapidly (e.g. 30% per metre) after leaving the laser and then expands again after a couple of metres, this implies that the output window is acting as a lens. In some cases the problem may be rectified by simply cleaning the output window, but it is generally best to replace it after establishing whether it was contaminated on the inside or outside. The source of contamination should then be identified and the problem rectified. Internal sources of oil contamination are the vacuum pump and oil coolants (if any). External sources include any vapours or dust. To exclude these it is usually advisable to have the external optical chain (particularly the output window) kept clean by a continuous supply of clean dry air. Compressor air will need to be filtered three or four times and should preferably be of breathable quality. "Raw" compressor air will do more harm than good as it contains large amounts of oil vapour and water.

6.2.10 External Mirror Condition and Alignment

Dirt can accumulate on the external mirrors, although it is common practice to purge the optical chain continuously with clean dry air to prevent this problem. As in the case of lenses, particles of dirt disrupt the beam in three ways:

1. They prevent the passage of light within their area.
2. They act as diffraction centres which can alter the mode.
3. They cause local heating by absorption which can distort or cause the failure of the optic.

Once again the problem is how to remove them without scratching the surface. The techniques are similar to those for cleaning lenses, but one specific method is worthy of note. As a mirror is a larger flatter surface than a lens, it is possible to clean them by laying a lens tissue over the whole surface. This tissue is then dampened with acetone or alcohol and removed from the mirror surface by gently dragging it across the surface from one side. In this way the mirror is wiped without any perpendicular force and scratching is minimised. The material from which the mirrors are made varies from manufacturer to manufacturer and some materials scratch more easily than others. Of the materials in common use, gold plated copper scratches most easily; molybdenum is the most scratch resistant. The slight drop in reflectivity between the two (gold ~99.5%, molybdenum ~98.5%) may be more than compensated for by the scratch resistance. Molybdenum is, however, rather an expensive material, compared with copper, and prices are therefore higher.

If the external mirrors are found to be out of alignment the procedures laid down in Chap. 5 should be followed. Checks for alignment should be made by taking beam prints at intervals along the optical path until the offending mirror is found (see Sect. 5.2).

Chapter 7

Safety Guidelines

7.1 Introduction

Carbon dioxide laser cutting machines have a good industrial safety record. Accidents are extremely rare if the machine is operational but the risks rise if maintenance work of any sort is carried out. The major danger to life is the high voltage system used to generate the electric discharge, and this should be worked on exclusively by well trained service personnel. This and other aspects of safety will be discussed under four main headings:

1. Beam exposure
2. Fumes
3. Electrocution
4. Fires

The following notes are offered as advice only and do not constitute instructions for safe working practice. Most countries now have CO_2 laser safety legislation and this should be studied carefully by owners of laser cutting machines.

Before the discussion begins, two extremely common misconceptions need to be clarified:

1. The word "radiation" which is often used when discussing lasers does not refer to atomic type radiation, it is merely another word for light. Carbon dioxide lasers do not emit any type of radiation except the light beam they are designed to generate.
2. Many people are under the impression that a clean energy source such as a laser will burn without pain and cut without bleeding. This is completely untrue. A surface skin burn (which usually turns the surface layer of skin black or in more minor cases, white) may not cause much pain at the time of burning but will start to hurt within minutes and will eventually ache like any other burn. If you are unfortunate enough to drill a hole or cut yourself on a focused laser beam (see Fig. 7.1) you will find it extraordinarily painful and must beware of delayed shock which might affect you several hours later. Laser cuts and burns do, however, heal rapidly and well compared with other physical damage as no mechanical force or chemical contamination is associated with them.

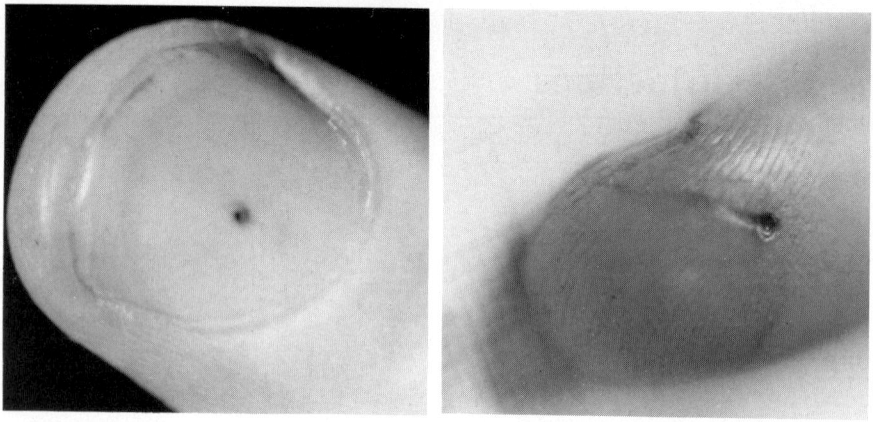

a b

Fig. 7.1. The author's finger after a very bad day. I was attaching a piece of paper to the nozzle for alignment (see Sect. 5.4.3) and the beam was inadvertently fired. **a** The neat circular hole in the nail is the exit; **b** the entry hole was elongated when I (understandably) removed my finger. (Photo taken one week after the accident.)

7.2 Beam Exposure

7.2.1 Skin Damage

The 10.6 μm wavelength light generated by CO_2 lasers is extremely effectively absorbed by water and hydrocarbon molecules. As the surface (and most of the bulk) of human beings is entirely made up of water and hydrocarbons (including most clothing) a person can be assumed to be an excellent absorber of CO_2 laser radiation. According to the British Safety Standard (BS 4803:83) the maximum safe energy density to which a human being can be exposed is 0.1 W/cm^2. A typical unfocused beam from a CO_2 laser (500 W with a diameter of 12 mm) has an average energy density of 442 W/cm^2 and the energy gradient within the beam is responsible for much higher energy densities towards the centre. This obviously implies that exposure to an unfocused beam for even a very short time will result in tissue damage. Exposure to such a beam for more than a few milliseconds may result in extremely serious burns. It is for this reason that the optical chain delivering the laser beam to the workpiece is fully enclosed. If the covers or "flight tubes" are removed from the optical chain a potentially dangerous situation is inevitable. A major feature of the hazard is the fact that the beam is infrared and therefore invisible. Exercise great caution when working around an unenclosed beam. Keep the number of people in the area down to a minimum, but do not work alone; someone must be close to a "beam off" switch at all times with instructions to switch the beam off if requested or if anything at all unexpected happens. There is of course no substitute for correct training, and this should be obtained from the equipment supplier whenever possible.

One source of beam exposure burns is reflection of the beam at or near the workpiece. Reflection from a cut zone is not a problem unless cutting is interrupted whilst the laser is still firing, e.g. if the oxygen supply fails, or the sheet material moves out of the focus zone – this can happen by the accumulation of thermal stress which will cause the sheet to distort suddenly.

In many cases the cut components fall vertically away from the parent sheet after cutting. During this fall or whilst they lie at the bottom of the cutting table at various angles these cut pieces can give rise to potentially dangerous reflections.

Another source of danger is reflections off the jigs used to hold work-pieces. Jigs are sometimes made from aluminium or brass as these materials are easy to machine. These materials are very reflective and care must be taken to avoid stray reflections off them.

First aid knowledge and equipment relevant to burns and shock is very useful and should be available as training and on wall charts near the laser installation. In all but the most minor superficial cases medical aid should be sought immediately after an accident. It is important that the person hurt in the accident is not allowed to drive and is kept accompanied, as the symptoms of shock can remain hidden for several hours.

7.2.2 Eye Safety

7.2.2.1 CO_2 Lasers

When discussing other types of laser (e.g. argon ion or helium–neon), eye safety is dealt with as an extremely important special case. These visible laser beams are focused by the eye down to a very high energy density within the eye itself and cause great optical damage. The situation with CO_2 lasers is slightly different, but no less dangerous.

Eye damage from a CO_2 laser will be primarily a surface effect rather than one of internal damage. It has already been mentioned that CO_2 laser light is readily absorbed by water and hydrocarbon molecules. A CO_2 laser beam impinging on the surface of the eye will be heavily absorbed by the surface layer of water and will rapidly dehydrate that layer and begin damaging the material beneath.

It is therefore necessary to protect the eye by using safety goggles of which many varieties are available ranging from simple plastic (polycarbonate) spectacles to high technology multilayer goggles which can withstand high energy densities for several seconds. It is important to note at this point that safety glasses suitable for CO_2 lasers will probably not give any protection from eye damage by other lasers. This is particularly important when using Nd:YAG or HeNe lasers. Certain multilayer goggles do, however, cope with a variety of laser types.

7.2.2.2 HeNe Lasers

It was mentioned in the alignment notes of Chap. 5 that many laser cutting machines are fitted with helium–neon (HeNe) lasers as alignment devices. The beam from these lasers is in the visible red part of the electromagnetic spectrum and is therefore a serious potential eye hazard. The output power of these alignment lasers is usually in the milliwatt range and beam diameters are typically 1–5 mm. At these power densities the beam is not capable of damaging skin tissue. On the other hand, severe optical damage can be experienced by the eye. The lens in the eye focuses the incident light very effectively onto the retina. The focused spot thus produced can have an intensity which is 100 000 times greater than the incident beam. Under these conditions the retina can become badly damaged.

The obvious instruction to any worker is therefore *never* to look down the beam. This instruction is particularly important for people experiencing lasers for the first time. The red HeNe laser light is extremely attractive and has a speckled appearance; the temptation to look down towards its source is rather high amongst newcomers or visitors to an installation.

Reflections of the HeNe beam off the workpiece, etc., are generally less hazardous than the original beam, but care should be taken when working with polished surface metals. Helium–neon beams are often used as a method of checking the proposed path of the CO_2 beam over the workpiece. The cutting or welding programme is executed using the HeNe beam rather than the high power CO_2 beam. The path of the red spot over the surface of the material will be identical to that of the CO_2 beam and positioning of the workpiece can be checked. If the movement is in two dimensions the workpiece surface will be perpendicular to the incident beam, and most of the reflected light will travel back along its original path. Viewing the red spot created on the workpiece is generally safe under these conditions.

If the path checking technique is being carried out for a three-dimensional profile, it may be possible to align the path of the incident beam accidentally with the inclined face of the workpiece and receive a strong direct reflection in the eye. Unfortunately, safety goggles for HeNe beams generally filter out all the HeNe light and so the spot you are trying to observe becomes invisible. Detailed safety information should be available from any supplier of such a system.

7.3 Fumes

All laser cutting devices generate fumes whilst cutting. These fumes are made up from the gases, liquids and particles ejected from the cut zone and are, in general, potentially harmful to health. For this reason all cutting machines should be fitted with an effective fume extraction system. No fume extraction is 100% effective, and for this reason a brief review of some of the most noxious and obnoxious fume generating materials is given here, together with relevant comments. For obvious reasons, fumes will be discussed under two headings: metals and non-metals.

7.3.1 Metals

7.3.1.1 General Hazards

Metal Fume Fever. This clinical condition has been recognised for centuries under such names as "brass founders ague". It is characterised by chills, fever, nausea, vomiting, muscular pains, dryness of the throat and mouth, fatigue, headache and weakness.

It usually develops about 3–6 h after exposure to fumes and normally disappears between 24 and 48 h. Recovery is said to be complete. The causative agents are the oxides of most metals particularly zinc, titanium, iron and tungsten. The most common source of this type of complaint to a laser operator using fume extraction will be zinc coated steel or titanium alloys.

Oil and Degreasing Agents. Oil is often used as a temporary protective coating on mild steel sheet in order to prevent corrosion. During cutting this oil can become vaporised, giving off fumes with a moderately unpleasant smell. These hydrocarbon fumes can be carcinogenic.

Some thick deposits of oil or grease can interfere with the cutting process as well as giving off potentially harmful fumes. It is not uncommon to remove this oily layer on the steel surface by wiping with a degreasing agent. These degreasing agents are commonly chlorinated hydrocarbons which are themselves a respiratory hazard, so necessary precautions (fume extraction etc.) need to be taken. The degreasing agent and its vapour are usually inflammable, and care should be taken not to allow these materials to get too close to the laser cutting table.

7.3.1.2 Specific Hazards

The individual metals mentioned in the following section are listed alphabetically. This list gives a general overview of a number of specific hazards, but is not exhaustive (i.e. the absence of a metal from the list does not imply that the fumes associated with it are harmless).

The Toxic Limit Values (TLVs) given here are a measure of the maximum permitted airborne concentration of fumes or dust for safe working. These are also known as Time Weighted Averages (TWAs), and a more complete list is available from the British Health and Safety Executive (Guidance Note EM 40/87).

Beryllium (TLV = $0.002\,mg/m^3$). A common alloying addition in alloys of aluminium, magnesium and copper. Can cause delayed serious damage to the lungs which, in the worst cases, can be fatal (note very low TLV).

Cadmium (TLV = $0.05\,mg/m^3$). The most common source of cadmium oxide poisoning is cadmium plated steel, which is sometimes used in preference to zinc plated steel. Cadmium oxide is extremely toxic, a powerful respiratory irritant which can lead to emphysema as a chronic condition. If exposure is over a prolonged period of time it can cause permanent kidney and liver disease. (Cadmium oxide poisoning is sometimes misdiagnosed as the less serious metal fume fever.)

Cobalt (TLV = $0.1\,mg/m^3$). Often present in stainless steel, the oxide can be a lung irritant.

Copper (TLV = $0.2\,mg/m^3$). See notes on metal fume fever.

Iron (TLV = 5.0 mg/m^3). This is probably the most common fume contaminant of the laser cutting environment. It is not necessarily a hazard, although it can accumulate in the lungs. Excessive exposure can result in metal fume fever.

Lead (TLV = 0.15 mg/m^3). Cutting of lead coated or lead containing materials may generate harmful lead fume levels. Lead poisoning is systemic, i.e. it enters the body via the lungs, but causes damage throughout the body with the exception of the lungs. The most common areas of affliction are:

The blood – causes anaemia by destroying red blood cells.
The gut – causes colic and/or constipation.
The nervous system – causes complaints from mania to depression.

Magnesium (TLV = 10.0 mg/m^3). A common addition to aluminium alloys; see notes on metal fume fever.

Manganese (TLV = 5.0 mg/m^3). A common alloying addition in stainless steels. In high concentration it can be a very nasty toxic material, acting as a respiratory irritant which causes high levels of inflammation (pneumonitis).

Nickel (TLV = 1.0 mg/m^3). A common alloying addition in stainless steels. Can act as a respiratory irritant and has been implicated as a lung carcinogen.

Titanium (TLV = 10.0 mg/m^3). See notes on metal fume fever.

Vanadium (TLV = 0.05 mg/m^3). Vanadium pentoxide is a highly irritant fume. Effects are mainly local to the eyes, throat and respiratory tract.

Zinc (TLV = 5.0 mg/m^3). See notes on metal fume fever. Zinc oxide fumes are most commonly generated by zinc coated (galvanised etc.) steel sheet.

7.3.2 Non-metals

7.3.2.1 Polymers

Polymers are generally made up of large, complex hydrocarbon molecules the degradation products of which range from harmless (water, CO_2 etc.) to extremely toxic (hydrogen chloride (HCl), hydrogen cyanide (HCN) etc.). Many of the vapour phase molecules given off during cutting are carcinogenic and fume extraction should be employed in all cases. There are some polymers which give off fumes which cannot be dealt with by simple extraction. The most common of these is polyvinyl chloride (PVC). When PVC is laser cut the molecule breaks down to HCl which is extremely toxic. As the gas comes into contact with living tissue it combines with any available water to become hydrochloric acid which then destroys the surrounding tissue. If ammonia is mixed with the air jet used to blow away the vapour from the cut zone the HCl will react with it to become the relatively harmless ammonium chloride. This techniques works as long as the balance between HCl evolution and ammonia supply is correctly controlled, but is still intrinsically dangerous. Other materials worthy of special attention are polyurethane, which is reputed to give off hydrogen cyanide, and fluorinated hydrocarbons such as polytetrafluoroethylene (PTFE), which may degrade to hydrogen fluoride. The cutting of many thermoset materials often generates a choking acrid dust which has not been identified as yet but which is undoubtedly harmful to health.

The basic rule of fume control for laser cutting should be that if you can clearly smell the fumes then the extraction system should be improved.

7.3.2.2 *Wood Based Products*

The burning process in the cut zone is little different to that experienced in a wood burning fire. The fumes given off untreated wood are therefore of little danger to health. Materials such as plywood and paper are given their strength by the addition of polymeric components, and these can cause the same problems as they do when cut alone.

7.3.2.3 *Ceramics and Glasses*

During laser processing of ceramics and glasses a very fine, hard dust of ceramic particles is generated which could be a respiratory irritant. Long term exposure could lead to forms of silicosis. Working in this gritty atmosphere can also give rise to sore eyes as a result of abrasion from particles caught on the eye surface. This problem may be accentuated if contact lenses are worn.

7.4 Electrocution

Throughout the entire history of lasers in all their forms there have been very few fatal accidents. These fatalities have all been the result of electrocution, generally of very experienced people working unprotected on the high voltage laser power supply. The potential dangers of working inside the laser or its power supply cannot be overstated. The only safe course of action is to call in a service engineer from the equipment supplier if a problem arises.

The covers and doors of the various parts of the laser are generally interlocked to prevent operation of the laser or its high voltage supply if they are removed. It is of course possible to defeat these interlocks, but this automatically results in a potentially lethal situation. Even if the laser has been disconnected from its electrical supply, there are hazards associated with the large capacitors found in many systems. All potentially charged components need to be earthed, even if the machine has been disconnected for hours.

All personnel associated with running the laser machine should be forbidden to open any of the interlocked access parts to the laser and its power supply unless they have been fully trained by the manufacturer. In the case of a trained engineer working inside the machine, an assistant should be on hand who is trained in how to deal with severe electric shock and methods of resuscitation. Information wall charts on how to deal with an electrocuted person should be available close to the laser itself and practice sessions held for the relevant staff.

7.5 Fires

The laser cutting environment is generally associated with sparks, hot vapours etc., and can be considered to be a moderate fire risk. As well as fires started by hot material leaving the cut there is a danger of ignition of the surroundings as a result of the reflection of the beam or mispositioning during alignment. To avoid fires around the cutting head the number of components made from non-metallic materials should be kept to a minimum. Extraction system tubing and gas pipelines bringing cutting gas to the nozzle should be either made of metal at the end nearest the cutting head, or shielded by a metal liner of some sort. Cables in this area should also be shielded.

This "absence of inflammable materials" philosophy should be extended to the whole area around the laser workplace. Small minor fires can become commonplace if the area around the cutting table is allowed to become untidy with worksheets, notepads and cleaning cloths.

Fire extinguishers for all types of fire should be available in the work place, together with relevant training in their use. Personnel must be trained in general about how to cope with fires, even if this training merely takes the form of an instruction to ring the fire alarm and leave the area by a certain route. One important point specific to the high power laser environment is that it is advisable to have a "laser shut down" control completely remote from the machine itself. In case of emergency the laser can therefore be deactivated without risk.

Chapter 8

Alternative Cutting Methods

The following notes are intended as a quick reference guide to profiling techniques which compete with, or complement, CO_2 laser cutting. The subject of cutting covers an enormous range, much of which is not relevant to our present discussion. The following sections will describe various features of some of the main industrial profiling techniques which can be accurately controlled by CNC methods.

1. Nd:YAG laser cutting
2. Plasma arc cutting
3. Abrasive water jet cutting
4. Oxygen–flame cutting

8.1 Nd:YAG Laser Cutting

Nd:YAG lasers, usually referred to simply as YAG lasers, are capable of cutting and welding metals in much the same way as CO_2 lasers. The output beam is generally of a lower power than that generated by CO_2 machines and this tends to restrict cutting speeds and maximum material thicknesses. The light generated by an Nd:YAG laser has a wavelength of 1.06 μm which is, by coincidence, one tenth of the wavelength of CO_2 lasers. This shorter wavelength light has a number of advantages over CO_2 laser output:

1. Glass is transparent at this wavelength, and so high quality glass optics can be employed to focus the beam. Glass lenses are generally of a higher optical quality than zinc selenide or gallium arsenide lenses used for CO_2 lasers and are also cheaper. Complex multicomponent optical arrangements can be employed to minimise the focus spot diameter, and it is possible to fit an optical microscope to the focusing head to observe the welding or cutting process directly through the same lens which is focusing the laser beam. In addition to these advantages it is possible to guide the beam down optical fibres, a subject which will be discussed later in this section.
2. A more intense focused spot can be achieved than is possible for CO_2 lasers. The diameter of a focused spot is proportional to the wavelength of the light being focused. As the Nd:YAG laser generates light with a

wavelength of 1.06 μm rather than 10.6 μm the minimum focal spot diameter will be approximately one tenth of the CO_2 value.

3. The absorptivity of metals to infrared light increases as the wavelength decreases. The increase in absorptivity for the shorter wavelength Nd:YAG laser implies that these machines would be suitable for processing highly reflective materials. In practice this has been found to be true, and Nd:YAG machines have found applications in the jewellery industry, cutting gold sheet.

In spite of all these optical advantages, Nd:YAG lasers occupy only a small part of the general laser cutting market. The majority of Nd:YAG machines are sold for welding and drilling applications, which is a reversal of the CO_2 laser market. The reasons why CO_2 lasers dominate the field are as follows:

1. Although cutting speeds for most metals are similar for both Nd:YAG and CO_2 machines at any particular laser power, the usual range of powers for Nd:YAG cutting machines is much lower (50–400 W) than it is for CO_2 machines (50–2000 W). At the highest powers (⩾400 W) the mode of Nd:YAG lasers becomes corrupted and this undermines the optical advantages as the beam becomes difficult to focus.
2. A major part of the cutting market involves the profiling of polymers. Carbon dioxide laser radiation is very effectively absorbed by these materials, but they are generally transparent, and therefore uncuttable at the Nd:YAG wavelength of 1.06 μm.

Nd:YAG lasers have found a market in applications where fine detailed work or highly reflective metals need to be cut. For this reason, Nd:YAG cutting machines generally have an *XY* movement of a few centimetres, rather than the usual CO_2 stroke length of up to 3 m. Kerf widths of tens of micrometres can be achieved, and dross is not generally a problem as the amount of residual melt left on the cut edge is small. Cutting takes place as a result of repeated pulsing of the beam. Although the laser power is quoted in watts, this refers to the average power over several seconds. A 400 W laser for example pulses at frequencies of up to 500 Hz (pulses per second) with each pulse achieving a peak power of 5–6 kW.

The utilisation of optical fibres for delivering the beam to the workpiece has been developed in recent years and has begun to find commercial application. The advantages of such a system are obvious where robotic arms are already in service. The laser fibre can easily replace or accompany another tool on such a device. The laser hardware can be kept well clear of the working zone, and the cutting head on the end of the fibre can be made as small as required. The major disadvantage is that the output mode from fibres is very poor when compared with the original output beam of the laser. This mode corruption gives rise to larger kerf widths (e.g. 0.5 mm) and lower cutting speeds, but the cut edge quality remains acceptable. Optical fibres are now being used on car production lines to carry out cutting jobs such as the trepanning of holes for fixtures.

Although Nd:YAG lasers cannot cut polymeric materials, they can be used to cut ceramics with great success by scribing or full penetration cutting. The maximum speeds of scribing are lower than for CO_2 machines

Table 8.1. ND:YAG Cutting results (Refs 69, 76 and 78, Chap. 11)

Material[a]	Thickness mm	Thickness in	Average laser power	Cutting speed m/min	Cutting speed in/min	Pulse Frequency Hz	Pulse Length ms
Mild steel	2.5	0.1	350	0.559	22.0		
	5.0	0.2	350	0.127	5.0		
	10.0	0.4	350	0.010	0.4		
	10.0	0.4	500	0.10	4.0		
Stainless steel	0.5	0.02	120	1.0	40.0	100	0.5
	2.0	0.08	120	0.45	18.0	60	0.5
	4.0	0.16	120	0.1	4.0	30	1.5
	1.0	0.04	400	0.9	36.0		
	3.0	0.12	400	0.5	20.0		
	5.0	0.2	400	0.25	10.0		
	10.0	0.4	400	0.1	4.0		
	1.0	0.04	500	3.0	120.0		
	2.5	0.1	500	0.8	32.0		
	10.0	0.4	500	0.16	6.4		
	15.0	0.6	500	0.07	2.8		
Aluminium	1.0	0.04	120	0.5	20.0	100	0.5
	3.0	0.12	120	0.05	2.0	30	1.5
	2.0	0.08	500	0.75	30.0		
	6.3	0.25	400	0.1	4.0		
Copper	1.0	0.04	120	0.5	20.0	60	0.2
	3.0	0.12	120	0.05	2.0	15	2.00
Titanium	1.0	0.04	120	1.0	40.0	100	0.5
	3.0	0.12	120	0.3	12.0	30	1.5
	4.0	0.16	400	0.25	10.0		

[a] Cutting gas is oxygen for all materials except titanium, where argon is used.

because the pulse repetition rate is lower for Nd:YAG (i.e. maximum ~500 Hz). Some users have found that the quality of the cuts is superior as the lower depth of residual melt left on the cut edge can be easily removed by shot blasting. The thicker layer of residual melt associated with CO_2 laser cutting is not easy to remove and can be a source of potential mechanical failure of the cut component. Table 8.1 gives a selection of cutting speeds for various materials using Nd:YAG lasers.

8.1.1 Principle of Operation of Nd:YAG Lasers

Nd:YAG lasers differ from CO_2 lasers in that the lasing medium is solid, and the lasing transition involves changes in the internal structure of an ion rather than the mechanical distortion of a molecule.

Figure 8.1 shows a schematic of an Nd:YAG laser. The lasing medium itself is a transparent rod of YAG (Yttrium–aluminium–garnet) which contains neodymium ions (Nd^{3+}). The neodymium ions are energised by the light from high power krypton or xenon flash lamps. This incoming energy raises the energy of Nd^{3+} ions by exciting one of the electrons orbiting the nucleus into a more energetic orbit. The ion stays in this metastable excited state until it is triggered to release its stored energy by an incident photon, which was itself released from just such an excited ion. Thus the lasing mechanism,

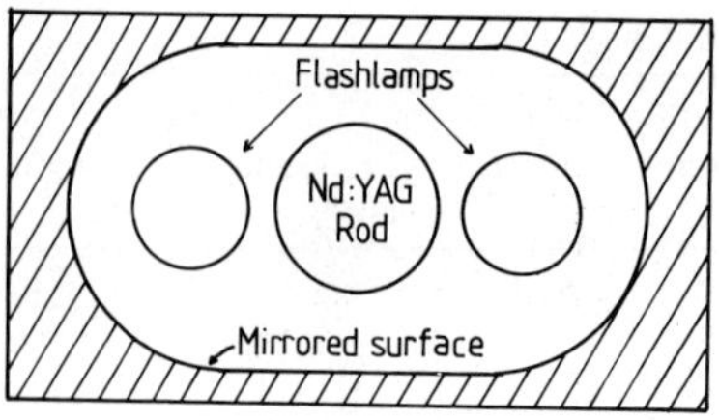

Fig. 8.1. A schematic diagram of an Nd:YAG laser shown in cross section. Laser light is emitted from the end of the Nd:YAG rod when it is energised by light from the high power flash lamps.

1 photon + 1 excited particle (ion) = 2 photons + 1 relaxed particle

which was mentioned in Chap. 1 and, in more detail in Chap. 9, remains unchanged whether the lasing particle is an ion or molecule. The mirrors at either end of the lasing cavity perform the same service as they do for CO_2 lasers. The YAG rod does not actually take part in the lasing process, but merely acts to hold the Nd^{3+} ions in place in the laser cavity where they can be illuminated by the flashlamps and subsequently lase.

This laser system can be used in the continuous wave (c.w.) mode if the flash lamps are left turned on, but output powers are low. The more energetic pulsed mode is favoured in materials processing applications. Although the switched nature of the flash lamps themselves gives rise to pulses of laser light, these output pulses can be made sharper and therefore of higher peak energy, if the laser is Q-switched. Q-switching involves the suppression of the lasing action until the maximum number of ions are in the upper lasing level. Lasing is then allowed to take place, and the enormous avalanche of lasing actions which ensues generates the maximum number of photons in the minimum time. This sudden very high pulse of energy can be achieved by exposing the Nd:YAG rod to a flash of light with one of the laser mirrors blocked off from the laser cavity by a shutter. When the shutter is then removed to expose the mirror the lasing action can start, and will rapidly de-excite the ions at the same time as releasing all the stored photons of laser light.

8.2 Plasma Arc Cutting

The heat source for plasma cutting is a high temperature, high velocity stream of partially ionised gas. This plasma jet is produced by constricting a welding type electric arc by passing it through a water cooled copper nozzle. Figure 8.2 shows the basic principle of action of a plasma jet in two modes:

1. The transferred mode, where the power supply is connected between the electrode and the workpiece.
2. The non-transferred mode, where the power supply is connected between the electrode and the nozzle.

The transferred mode is usually the one used as a cutting device as it provides a more efficient use of the arcs energy. The fundamental process parameters which control the condition of the arc, and therefore its effec-

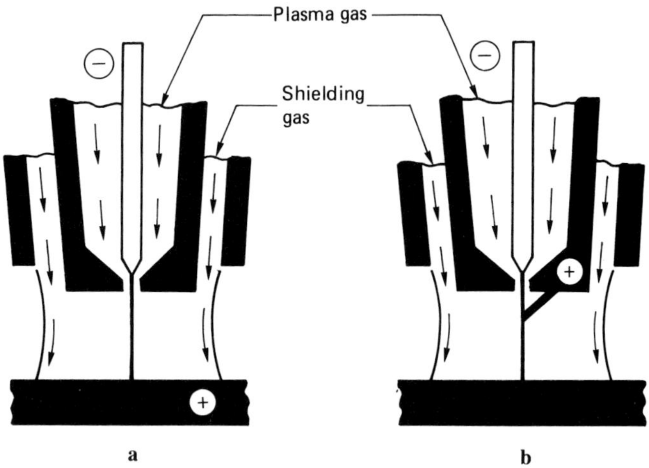

Fig. 8.2. A schematic of the principle of operation of plasma arc devices in: **a** the transferred mode; **b** the non-transferred mode.

tiveness, as a cutting device are nozzle size, arc current, gas type and flow rate. A heavy duty mechanised torch with a current capacity of up to 1000 A can cut through 150 mm thick stainless steel and 160 mm thick aluminium (Ref. 73, Chap. 11). In most applications, however, the plate thickness seldom exceeds 65 mm. The process is not generally used to cut mild or carbon steels in sections above 50 mm because at this thickness, oxygen–flame cutting is considered a cheaper alternative (Ref. 77, Chap. 11). The temperatures in the arc can be as high as 20–50 000 °C close to the electrode. A central column of high energy ionised gas is surrounded by a cooler boundary layer. The gas jet on many commercial devices incorporates a swirling action which tends to force the cooler un-ionised gas rapidly outwards, and helps maintain the high temperatures in the central core.

There are three different types of plasma torch:

1. Dual gas: one gas creates the plasma and the second provides an outer sheath which helps maintain the narrow central core of the plasma by protecting it from the shearing effects of the atmosphere. Common mixtures are argon (sheath)–hydrogen (plasma), argon–nitrogen and nitrogen–carbon dioxide.
2. Air plasma: in these devices compressed air is the only gas employed and running costs are minimised.
3. Water injection: this is basically the same principle as the dual gas system with water acting as the sheath rather than a protective gas.

The highest temperatures are achieved by water injection plasma devices which incorporate a radial shroud of water to help constrict the plasma and so maximise temperatures. Despite these extremely high temperatures of the arc, only one tenth of the water is vaporised. The remaining 90% of the water exits from the nozzle and acts as a coolant for the workpiece. This cooling, similar to water assisted laser cutting (Sect. 2.1.5.2), reduces the

Table 8.2. Typical cutting speeds for plasma arc cutting (Ref. 77, Chap. 11)

Material	Thickness mm	Thickness in	Cutting speed m/min	Cutting speed in/min
Mild steel	6	0.25	5.0	200
	13	0.50	2.5	100
	25	1.00	1.2	50
	50	2.00	0.6	25
Stainless steel	6	0.25	5.0	200
	13	0.50	2.5	100
	25	1.00	1.2	50
	50	2.00	0.5	20
	100	4.00	0.2	8
Aluminium	6	0.25	7.5	300
	13	0.50	5.0	200
	25	1.00	2.2	90
	50	2.00	0.5	20
	100	4.00	0.3	12

extent of the heat affected zones and the level of residual surface oxidation. The steam boundary layer around the arc helps to protect the nozzle from direct impingement of the arc, and nozzle lifetimes are improved.

The plasma arc technique can be used to cut most metals and this is particularly effective on thick sections. Table 8.2 shows the level of cutting speeds which can be expected from this process. In view of the variety of techniques possible, these results should be taken as an approximate guide only.

Although cutting speeds are high and maximum thicknesses are very much higher than for CO_2 laser cutting, plasma cutting suffers from the following drawbacks:

1. The kerf width is considerable (several millimetres) and is therefore a limit on the level of detail that can be cut.
2. The edges of the cut are bevelled. Typically this bevelling can be of the order of 2° to 3° or more from the vertical (see Fig. 8.3). As a result of this a 25 mm (1 in) thick plate could have a kerf width of 7.0 mm at the top of the cut and 4.5 mm at the bottom. In some cases this bevelling can be an advantage if the cut product is subsequently to be welded. In most cases, however, it would be preferred if the cut edges were perpendicular.

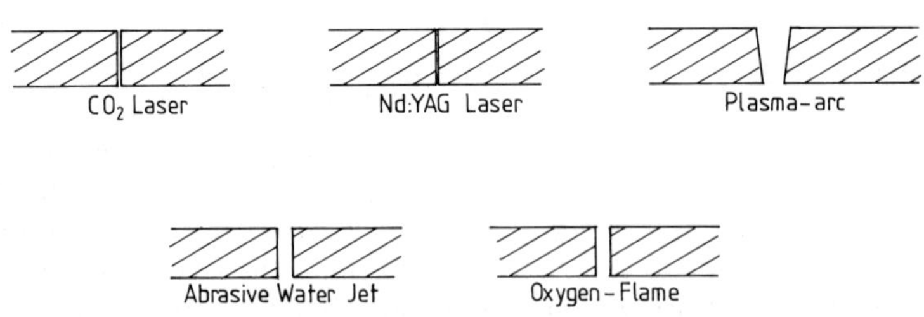

Fig. 8.3. A generalised comparison of the kerf cross sections produced by each process.

3. The electrical nature of the process generally restricts it to the cutting of metals.
4. The heat input to the workpiece is large and this results in extensive heat affected zones and thermal distortion on cooling.
5. Although the capital cost of a plasma arc gun may be an order of magnitude cheaper than a laser, the running costs are generally substantially higher.
6. The process is extremely noisy and, because of the high metal removal rate, generates a large quantity of dust and fumes.

In summary it can be said that plasma cutting is a very effective rough cutting operation for medium to thick sections. Cutting speeds are high for metals but the process does not lend itself to the profiling of non-metals. In cases where the cut components are to be welded, the bevel of the cut edge can be an advantage but in general the cut edge quality is the major restriction on the use of this technology.

Although the process is generally machine guided, it can be used as a hand held operation. In the case of manual manipulation it is very important that the nozzle does not touch the workpiece as this will earth the arc and cause substantial nozzle damage. For this reason the tip of the nozzle is often made of an insulating material.

Because the cost of the plasma arc cutting head is low compared with the price of *X-Y* manipulation equipment, it is common for a cutting table to have more than one cutting head. The cutting heads are supplied with electricity and gas through flexible umbilicals, and can therefore be mounted together on a moving or stationary beam above the cutting table. The distance between the heads can be adjustable and thus two or more heads can work together cutting components out of different parts of the same sheet.

Plasma cutting is a well established process in a wide range of industries from shipyard work to chemical plant. It also finds application in the forgings and castings industry for removing risers and gates. The process is best suited to non-precise cutting of medium section metals.

Micro-plasma. Recent advances in plasma cutting have resulted in the development of the micro-plasma process. This is a refinement of standard plasma cutting which gives smaller kerf widths and a more parallel sided cut. In spite of these improvements the process does not compete well with laser cutting because the kerf widths and start up (penetration) holes are still too large for the successful profiling of most bolt holes and other small details.

8.3 Abrasive Water Jet Cutting

In the early 1970s the first investigations into water jet cutting led to the development of very high pressure water jets capable of cutting a wide range of soft materials and non-metals. Approximately a decade later the process was extended to the cutting of metals with the advent of abrasive water jet cutting. Figure 8.4 shows a schematic of an abrasive water jet (AWJ) cutting system: A hydraulic pump is used to pressurise water in an intensifier to

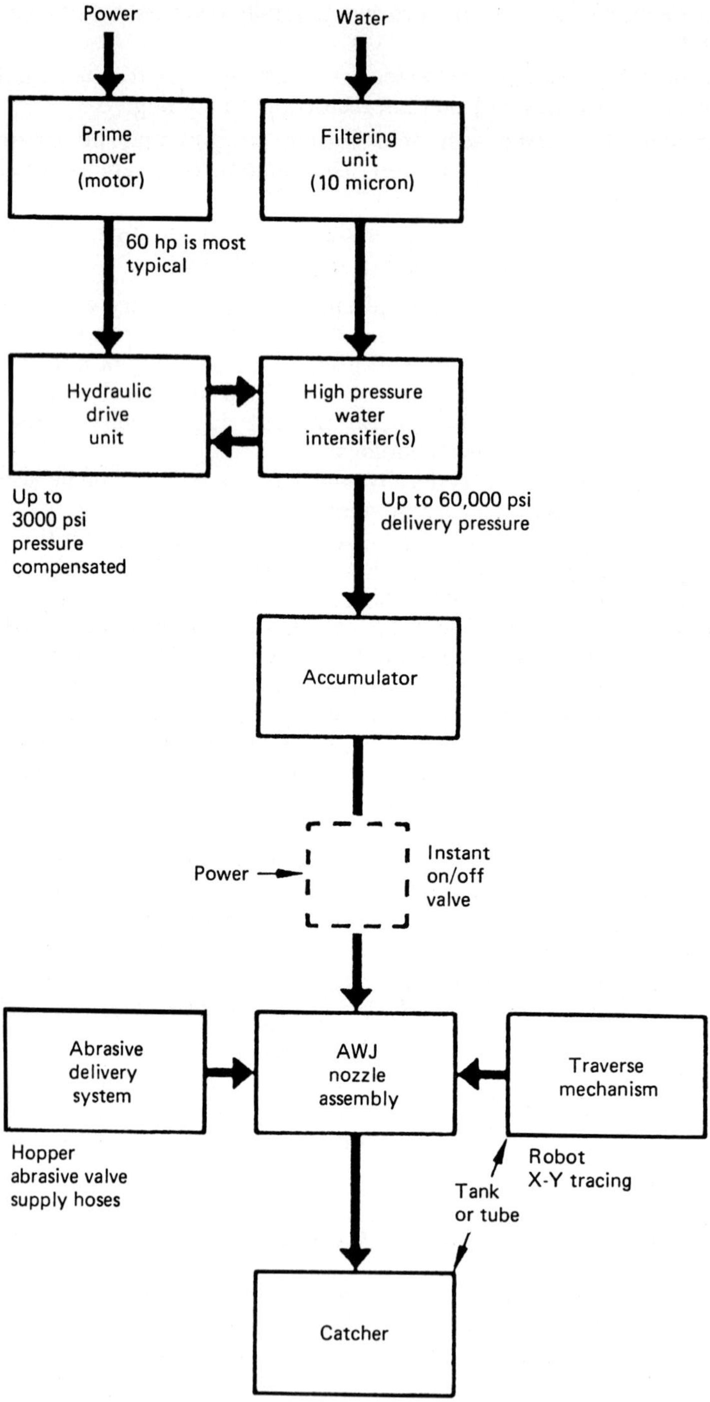

Fig. 8.4. A schematic of an abrasive water jet cutting system (Ref. 70, Chap. 11).

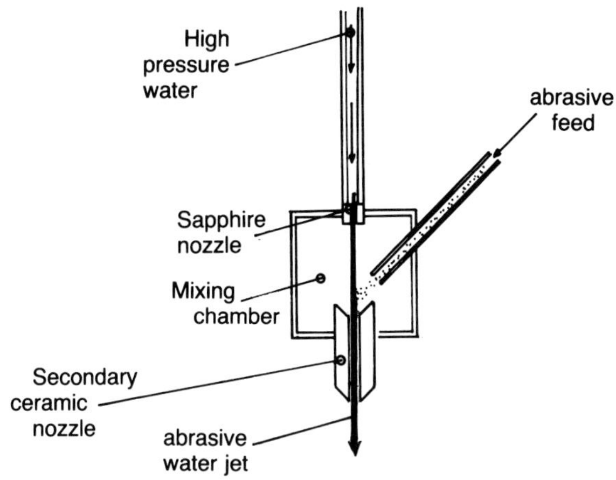

Fig. 8.5. A cross section of the abrasive water jet nozzle system demonstrating the mixing of the abrasive with the water.

pressures of the order of 50 000 psi. This pressurised water is then ejected from a small diameter nozzle. The high pressure water jet thus produced then passes through a chamber (see Fig. 8.5) where it is mixed with abrasive powder before passing through a final nozzle to interact with the workpiece. The intensifier achieves very high water pressure by using a large diameter, hydraulic oil driven piston to drive a smaller diameter piston acting on a chamber full of water. The pressure on the water exceeds the pressure of the oil by the ratio of the surface areas of two cylinders. This ratio can be of the order of 20:1 and thus a 2000 psi hydraulic oil system can produce a pressure of 40 000 psi in the water.

These intensifiers have a predetermined stroke length and are often designed to be reciprocating to maintain as continuous a supply of pressurised water as possible. Even though this reciprocating action produces pressurised water when the piston is travelling in either direction, the flow is not continuous as the piston must stop between changes of direction. To reduce the consequences of this intermittent flow on the cutting action of the pressurised water, the system includes an accumulator to act as a reservoir of high pressure water which then supplies the cutting nozzle. The abrasive can be mixed into the water by injecting a blown powder as shown in Fig. 8.5 or by filling the mixing chamber with a water and abrasive slurry through which the high pressure jet can pass.

The actual cutting mechanism is one of erosion of the workpiece by the high velocity hard particles in the water jet. These particles, which are commonly made of silica, garnet or aluminium oxide, are generally used in grit sizes of 60 to 120 with consumption rates of 250 g/min to 2.0 kg/min. Fine abrasives of the mesh 120 type may be used when the best surface finish is required, and have been used with excellent results when cutting glass. Larger grit sizes (60 mesh; 250 μm diameter) are more effective for steel cutting, and even coarser sizes (e.g. mesh 36 and 16) have given the best results for cutting rocks and concrete (Ref. 71, Chap. 11).

Table 8.3 gives an idea of the cutting speeds available by this process. These figures must be used as a guideline only as the process involves a large

Table 8.3. Typical cutting speeds for the abrasive water jet method (various sources including Refs 70–72, Chap. 11)

Material	Thickness mm	Thickness in	Cutting speed m/min	Cutting speed in/min
Mild steel	1.6	0.063	0.50	20.0
	13.0	0.5	0.10	4.0
	50.0	2.0	0.038	1.5
	180.0	7.0	0.010	0.125
Stainless steel	5.0	0.2	0.40	15.0
	13.0	0.5	0.15	6.0
	25.0	1.0	0.076	3.0
Aluminium	1.6	0.063	1.30	50.0
	6.0	0.25	0.50	20.0
	25.0	1.0	0.13	5.0
	100.0	4.0	0.025	1.0
Titanium	3.0	0.125	0.50	20.0
	6.0	0.25	0.40	15.0
	12.0	0.50	0.10	5.0
Glass	13.0	0.5	1.3	50.0
	19.0	0.75	0.6	25.0
	25.0	1.00	0.13	5.0
Marble	50.0	2.00	0.4	16.0
Concrete	250.0	10.00	0.025	1.0

Fig. 8.6. Abrasive water jet cutting of 63 mm (2.5 in) thick steel. (Courtesy of Flow Industries Inc., Kent, WA, USA: Ref. 70, Chap. 11.)

number of independent and interrelated variables all of which have an effect on the effectiveness of the process. These parameters include water pressure, nozzle diameters, grit type and size etc. Figure 8.6 is a photograph of AWJ cutting in action, cutting 63 mm (2.5 in) thick mild steel (Ref. 70, Chap. 11). Inside the water jet the abrasive grit particles have been accelerated to velocities of the order of 500 m/s (1500 ft/s). At this velocity the impact of the hard particles on the cutting front causes local failure of the workpiece and controlled erosion takes place. Because high particle velocities are needed over the whole of the cutting front, the velocity and momentum of the water jet is only partially reduced during passage through the cut zone. It is therefore important to dissipate the high energy jet leaving the bottom of the cut zone. This is done either by catching the water in a hard ceramic tube long enough to disperse the jet, or for production systems where the jet is moving rather than the workpiece, an energy absorbing replaceable bed is used to support the workpiece (e.g. 1.5 m long wooden stakes stacked close together with their ends exposed to the jet or a 0.5 m bed of ball bearings).

The advantages of AWJ cutting over CO_2 laser cutting are:

1. The maximum thickness of material which can be cut is an order of magnitude or more than is possible by CO_2 laser. Steel cutting in thicknesses up to 100 mm (4 in) is common. Most materials can be cut in sections of several centimetres.
2. All materials can be cut irrespective of such physical properties as melting point and thermal conductivity. Materials which are impractical to cut by thermal methods such as marble or concrete are cut with great effectiveness.
3. As the process is non-thermal there is no heat affected zone associated with the cut edge. This feature is particularly attractive when susceptible materials such as titanium alloys are being profiled.
4. For some metals the cut edge quality can be superior. Dross is not a problem.

The features of the AWJ process which compare badly with CO_2 laser cutting are:

1. Kerf widths are substantially larger than for CO_2 laser cutting and generally exceed 2 mm (0.08 in). This naturally restricts the amount of detail which can be cut.
2. The process is generally much slower for cutting metals in the range of thicknesses where the two compete (i.e. up to ~16 mm (0.6 in)). For example, for mild steel 1.6 mm (0.06 in) thick: 1200 W CO_2 laser, 9.0 m/min (360 in/min); AWJ, 0.5 m/min (20 in/min).
3. The capital and running costs of the two processes are of the same order of magnitude, but the cost per unit length of cut will be greater for the AWJ process when thin section metals are cut as a result of the reduced cutting speed.
4. AWJ cutting is much noisier and generally messier than CO_2 laser cutting.

In summary, the AWJ process is an excellent method of cutting thick section materials of all types, but cutting speeds are generally low and fine detail is not possible. AWJ is generally a complementary process to CO_2 laser cutting rather than a direct competitor. If a laser can do the job in question it will probably do it cheaper. When laser cutting is incapable of cutting the material AWJ takes over.

8.3.1 Abrasive-Free Fluid Jets

Using the same technology as AWJ cutting without the abrasive, water can be used to cut a wide variety of non-metals. Examples of cutting speeds are given in Table 8.4.

Although water can be used to cut foodstuffs, the water uptake of the cut edge can cause problems. Materials such as biscuit have been cut with great success using vegetable oil as the cutting jet medium. Many prepared foods contain vegetable oil or are subsequently cooked in it and so the cut edges are no different from the base material. Vegetable oil does not wet the food to the same extent as water and cut edge pickup is minimal.

Table 8.4. Cutting speeds for pure water jet cutting (Ref. 72, Chap. 11)

Material	Thickness		Cutting speed	
	mm	in	m/min	in/min
PVC	0.76	0.03	18.3	720
Polypropylene	2.03	0.08	3.7	145
Polycarbonate	2.16	0.85	1.2	47
Asbestos board	18.00	0.70	91.2	360
Corrugated cardboard	7.87	0.31	20.0	780
Leather	1.50	0.06	76.2	3000
Rubber	3.15	0.12	9.1	360
Formica	1.27	0.05	36.6	1440
Carpet	18.29	0.72	7.6	300
Glass–epoxy laminate	1.77	0.07	2.5	100

8.4 Oxygen–Flame Cutting

Oxygen–flame cutting is similar to laser cutting with oxygen in that an exothermic reaction between the metal and oxygen is initiated by a localised heat source. In the case of oxygen–flame cutting the heating of the metal to its ignition temperature in the oxygen jet is carried out by a flame which surrounds the central core of pure oxygen, as shown in Fig. 8.7. This "preheat flame" can be generated by any oxygen–fuel gas mixture but the two most commonly used fuel gases are acetylene and propane.

The process dates back to the early 1900s and has been in widespread use ever since. The cutting head is merely a nozzle of the type shown in Fig. 8.7 attached by hoses to the gas supply and is therefore suitable for hand held use as well as machine controlled cutting. Hand held cutting is generally used for inaccurate rough cutting applications, but more accurate work can be carried out on machines whose cutting path is controlled either by template following techniques or CNC control.

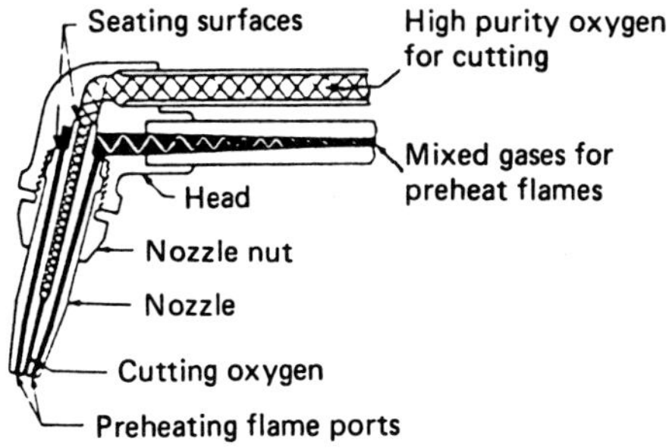

Fig. 8.7. A schematic of an oxygen–fuel gas cutting head (Ref. 77, Chap. 11).

The oxygen–flame cutting method is most commonly applied to mild steel, but some other exothermically oxidising metals can be cut as well. Table 8.5 gives some idea of cutting speeds possible by this method when cutting mild steel. The process is not, of course, suitable for the cutting of non-metallic materials. Comparing the figures in Table 8.5 with laser cutting speeds it is clear that although oxygen–flame cutting is a comparatively slow process at thin sections, it is capable of cutting much greater sections than are possible using a laser. Sections more than a metre thick can be cut but the quality will be poor and kerf widths very large (⩾10 mm). The process is not suitable for thin sections generally and kerf widths are always of the order of a few millimetres or more. If a large number of medium section (e.g. 6–12 mm (0.25–0.5 in)) components are to be cut it may be preferable to weld or clamp the workpieces together and cut them as a stack. The best tolerances are held up to a thickness of ~50 mm (2 in) with a gradual decay in quality up to 150 mm (6 in) where tolerances can only be held to ±1.5 mm (0.06 in) (Ref. 77, Chap. 11).

Cut surfaces are covered in striations which are similar to, but coarser than, those observed on laser cut specimens. The surface roughness of cut

Table 8.5. Typical cutting speeds for oxygen–flame cutting of mild steel (Ref. 77, Chap. 11)

Thickness		Cutting speed	
mm	in	m/min	in/min
3.2	0.125	0.4–0.8	16–32
6.4	0.25	0.4–0.6	16–26
12.7	0.50	0.3–0.6	12–23
19.0	0.75	0.3–0.5	12–21
25.4	1.0	0.2–0.4	9–18
38.1	1.5	0.15–0.35	6–14
50.8	2.0	0.1–0.3	6–13
101.6	4.0	0.1–0.25	4–10
152.4	6.0	0.07–0.2	3–7
203.2	8.0	0.07–0.1	3–5
254.0	10.0	0.05–0.1	2–4
305.0	12.0	0.05–0.1	2–4

NB: The highest speeds are for straight line cuts with only moderate quality. At the lower speeds the quality will be at its best (Ref. 77, Chap. 11)

edges becomes more pronounced as the thickness of the material increases. Heat Affected Zones (HAZs) on each side of the cut can be extensive particularly at thicker sections (e.g. ~5 mm HAZ (0.2 in) at a section of 150 mm (6 in) for a high carbon or alloy steel). Cut surfaces are generally approximately perpendicular to the cut sheet surface.

Oxygen–flame cutting does not lend itself to applications where the cutting process must be repeatedly stopped and restarted. The two reasons behind this are: (1) the flame and oxygen jet cannot be rapidly turned on and off with ease; (2) the flame and oxygen jet must initiate a cut in the middle of the sheet by blowing a hole in it. This process results in the explosive ejection of metal in all directions and this can block or damage the nozzle.

The major point in favour of oxygen flame cutting is the very low capital cost of the equipment, compared with a laser or most other cutting devices. On the other hand, the main drawback is the inability to carry out detailed work as a result of the large kerf width which is generally of the order of several millimetres. It is not surprising, therefore, that the majority of applications of oxygen flame cutting are restricted to rough profiling operations on medium section (6–50 mm (0.25–2.0 in)) section steels.

8.5 A Summary of the Strengths and Weaknesses of Each Process Compared with CO_2 Laser Cutting

Table 8.6 summarises the general conclusions of the earlier discussions by qualitatively comparing the four "alternative" cutting methods with CO_2 laser cutting. It is clear that the strongest points of CO_2 laser cutting are:

Table 8.6. A qualitative comparison of the four "alternative" cutting methods with CO_2 laser cutting

	Nd:YAG	Plasma	AWJ	Oxygen–flame
Capital costs	S	G	S	VG
Running costs	S	S	S	G
Metal cutting:				
Speed	P	G	VP	P
Quality	S–G	VP	VG	P
Maximum thickness	S	VG	VG	VG
HAZ	S–G	VP	VG	P
Kerf width	S–G	VP	VP	VP
Non-metal cutting:				
Ceramics				
Speed	P	N/A	P	N/A
Edge quality	S–G	N/A	G	N/A
Maximum thickness	S–G	N/A	VG	N/A
Kerf width	S–G	N/A	P	N/A
Polymers				
Speed	N/A	N/A	S	N/A
Quality	N/A	N/A	S	N/A
Maximum thickness	N/A	N/A	G	N/A
Kerf width	N/A	N/A	P	N/A

Key: N/A, not applicable; VG, very good; G, good; S, similar; P, poor; VP, very poor, compared with CO_2 laser cutting.

1. The range of materials which can be cut is very wide. In this respect the only superior cutting method is AWJ.
2. The kerf width is very small (0.1–1.0 mm), and thus the amount of detail which can be profiled is high. In this respect the only superior method is Nd:YAG laser cutting.
3. Cutting speeds for thin section metals (particularly steels) are generally high and are exceeded only by the plasma arc process. (Very highly reflective metals such as gold, silver and copper can be cut more effectively by high power Nd:YAG lasers.)

As Table 8.6 shows, the performance of CO_2 lasers is equalled or improved upon by more than one of the other techniques for some of the other aspects of cutting shown here. From amongst these comparisons it is clear that CO_2 laser cutting has two weak points when compared with the other techniques:

1. Capital costs are high. Although Nd:YAG and AWJ systems cost a similar amount of money, plasma and flame equipment is very much cheaper.
2. The maximum thickness of metal which can be cut is very low compared with all the other techniques except Nd:YAG machines. Although researchers have managed to cut 30–40 mm of steel in laboratories, the commercial maximum cutting thicknesses are approximately as follows:

Mild steel	18 mm (0.7 in)
Stainless steel	12 mm (0.5 in)
Aluminium	6 mm (0.25 in)
Copper	4 mm (0.16 in)

   These figures refer to the highest power commonly used for cutting (~2 kW). The more usual high power of ~1200 W will cut up to thicknesses of between half and two thirds of these values.

In spite of these two drawbacks, the CO_2 laser cutting industry has an assured future in profiling a wide range of materials. The quality and speed of the process has impressed an ever growing number of engineers and managers, and continuous advances in the technology are giving rise to new applications all the time.

Chapter 9

The Physics and Design of CO_2 Lasers

9.1 Introduction

Although the following description of the physics and design of CO_2 lasers will be rather more detailed than the earlier discussion in Chap. 1, it is not my intention suddenly to change style to the highly specialised language and mathematics usually associated with laser physics. Readers who require a high level of theoretical information will find that they are very well provided for in the large number of laser related physics textbooks which have proliferated over the past 20 years or so.

In common with many other aspects of modern physics, the fundamental interaction behind the generation of a laser beam was first postulated by Albert Einstein. In 1917, Einstein proposed that light of a particular frequency could, under the correct circumstances, stimulate atomic electrons to generate more light of the same frequency. More than 40 years later, in 1958, C.H. Townes and A.L. Shawlow proposed that the principles of microwave amplification by stimulated emission used to produce a maser could be extended to the amplification of light. A few years later (in 1960) T.H. Maiman built the first operating laser. The lasing medium was a ruby rod with mirrored ends which was excited by light from a high power flash lamp.

The story of the CO_2 laser begins with the pioneering experimental work carried out by C.K.N. Patel at the Bell laboratories in the USA in the early 1960s. Patel's first CO_2 laser, first reported in 1964, relied on the rather inefficient lasing mechanism of pure CO_2 and output powers were in the range of 1 mW. Within a year this output power had been raised to 10 W as a result of adding nitrogen to the CO_2. Subsequent increases in power and efficiency were achieved by the addition of helium to the mixture. By 1968, Patel was able to report (Ref. 55, Chap. 11) continuous wave powers of 80 W per metre of discharge tube and the existence of a CO_2 laser with an output of 8.8 kW.

Once it had been established that very high powers were available from CO_2 lasers, a large number of workers became employed in the task of making machines of improved efficiency, power and quality. Much of the design work has concentrated on two fundamental points:

1. Excitation of the gas
2. Withdrawal of surplus energy from the gas

Excitation studies have investigated direct current (DC), alternating current (AC), radio frequency (RF) and gas dynamic methods of providing the lasing gas mixture with energy. Surplus energy withdrawal investigations have generally concentrated on gas cooling by conduction to the walls of the laser cavity or by passage of the gas mixture through heat exchangers.

The following sections are intended to explain the physics of CO_2 lasers and various aspects of their design. Materials processing lasers have now diversified in design so much that it is possible to buy as standard items anything from a 4 W hand held device which will cut many non-metallic materials to a 10 kW machine capable of welding steel in sections of up to 12 mm (0.5 in). From the point of view of laser cutting however, it must be borne in mind that a higher power machine is not necessarily an improvement. The mode of the beam is a critical factor in determining the suitability (or otherwise) of any laser as a cutting device.

9.2 The Physics of CO_2 Lasers

9.2.1 The CO_2 Energy Spectrum

The CO_2 molecule is linear and symmetric with the two oxygen atoms attached to opposite sides of the central carbon. This molecule has three fundamental vibration modes which are:

1. *The symmetric stretch mode.* In this mode the oxygen atoms oscillate symmetrically along the axis of the molecule. The oxygen atoms move in unison away from, and then back towards, the central carbon atom.
2. *The bending mode.* In this mode the atoms of the molecule oscillate in a direction perpendicular to the axis of the molecule.
3. *The asymmetric stretch mode.* In this case the atoms oscillate along the axis of the molecule in an asymmetric manner. The oxygen atoms "take turns" in being closest to the central carbon atom.

These three bending modes are not mutually exclusive, and it is possible for a molecule to be executing all three at once. Each mode has a different specific energy associated with it which is determined by quantum mechanics. In addition to this, it is possible for a molecule to be excited to a state where it contains more than one quantum of energy in any or all of the above modes (e.g. a molecule could be in the state of experiencing symmetrical stretching, bending and a "double dose" of asymmetrical stretching). In order to identify the energetic state of a particular CO_2 molecule it is given a number code $[V_1\ V_2\ V_3]$. The numbers V_1 V_2 and V_3 relate to the numbers of vibrational quanta in each mode.

V_1 refers to the number of vibrational quanta in the symmetric stretch mode.

V_2 refers to the number of vibrational quanta in the bending mode.

V_3 refers to the number of vibrational quanta in the asymmetric stretch mode.

For example, a CO_2 molecule experiencing a single vibrational quantum in the bending mode and three quanta in the asymmetric stretch mode would be denoted as CO_2 [0 1 3].

9.2.2 Interactions in a Pure CO_2 Laser

Figure 9.1 shows the area of the energy spectrum of CO_2 which is of interest as far as high power lasers are concerned. The lasing process involves the release of stored energy in the form of photons of light. In order to generate this light the CO_2 molecule must relax from a highly energetic state (the upper lasing level) to a lower one (the lower lasing level). Using the V_1 V_2 V_3 terminology, CO_2 molecules release the 10.6 μm photons when relaxing from their [0 0 1] state to the lower energy [1 0 0] condition. In other words the upper lasing level consists of a single quantum asymmetric stretch and the lower lasing level is the single quantum symmetric stretch mode. In a pure CO_2 laser the molecules are directly excited to the [0 0 1] state by collision with high energy electrons. The molecules which eventually contribute to the lasing action remain in this high energy state until they are stimulated by an incident lasing photon. At this point the molecule rejects the incoming photon and simultaneously relaxes to the [1 0 0] state, releasing an identical "twin" of the incident photon. (See Sect. 9.2.5 on the lasing mechanism.)

The relaxation from [0 0 1] to [1 0 0] is not the only lasing reaction possible in a pure CO_2 laser. Decay from the [0 0 1] state to the [0 2 0] condition can take place to generate an output of 9.6 μm wavelength, but this effect is minor compared to the [0 0 1]–[1 0 0] transition generating 10.6 μm wavelength radiation. This [0 0 1]–[0 2 0] transition will be ignored in the interests of clarity.

The discussion so far has concentrated solely on vibrational energy states but this rather over simplifies the situation. As well as storing and releasing quanta of energy as a result of vibrational states, the CO_2 molecule has a number of rotational states within which energy is stored and released as a result of the twisting of the molecule. The amount of energy associated with

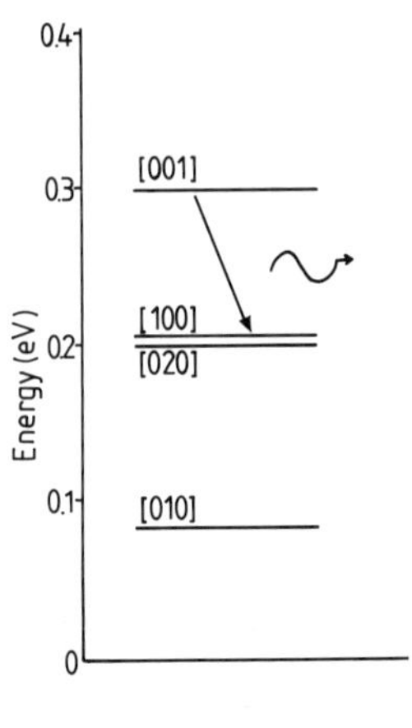

Fig. 9.1. A simplified schematic of the energy spectrum of the CO_2 molecule showing the transition from the [0 0 1] state to [1 0 0] which generates a photon of 10.6 μm wavelength light. (The molecule is initially excited to the upper state [0 0 1] by collision with high speed electrons.)

these rotational states is smaller than the vibrational energy levels. The rotational movement is independent of the vibrational state and so each vibrational state is divided up into various energy levels associated with the different levels of rotational energy possible. These rotational energy states have specific energies determined by the quantum mechanical selection rule which only allows those transitions involving a change in angular momentum of $\pm h/2$ (h is Planck's constant). When a CO_2 molecule is excited up to its vibrational upper laser level [0 0 1] it also assumes one of the rotational levels. There is therefore a distribution of energy within the [0 0 1] level with various proportions of the energised gas molecules at each of the rotational levels. During the relaxation process which generates a photon, the energised CO_2 molecule can, in theory, relax from any of a number of rotational energy levels within the [0 0 1] vibrational state to any of a similar number of rotational levels in the [1 0 0] state. This point is demonstrated in Fig. 9.2.

This process gives rise to a spectrum of possible output because the energy drop from the upper level to the lower level determines the energy of the emitted photon. This energy determines the wavelength of the photon:

$$E = h\nu$$

where

E = the difference in energy between the upper lasing level and the lower
h = Planck's constant
ν = the frequency of the light emitted (the wavelength of the light can easily be determined from the frequency:
$\nu = c/\lambda$ where λ is the wavelength and c the speed of light)

If all the various transitions from the [0 0 1] to the [1 0 0] state played a part in producing the output from a CO_2 laser the eventual beam would be made up of a number of wavelengths. The polychromatic nature of the beam would, unfortunately, undermine many of the advantages of a CO_2 laser as a materials processing device. In order to achieve the maximum energy density at the focused spot the beam must be as monochromatic

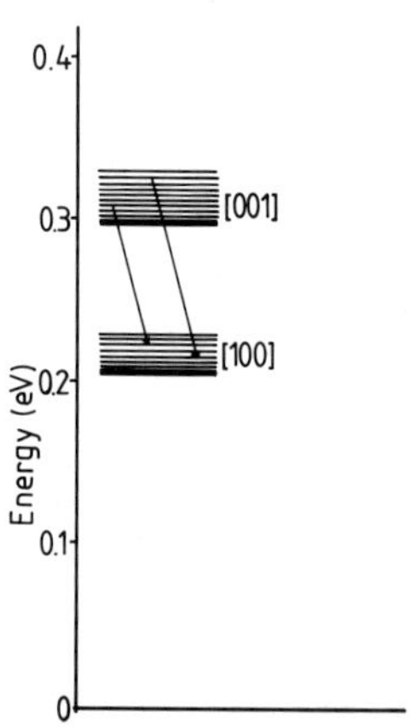

Fig. 9.2. A schematic showing the rotational levels in the upper and lower vibrational states. When an incident photon stimulates the molecule to relax from the upper vibrational state [0 0 1] to the lower [1 0 0] the transition may be from any of the [0 0 1] rotational states down to any of the [1 0 0] rotational states. This range of possible transition energies would give rise to a range of output wavelengths, but this effect is minimised by selective competition between the transitions.

(single wavelength) as possible. It is fortunate, therefore, that selective competition between the numerous possible energy transitions results in the domination of one transition which therefore results in the great majority of the output photons being of a single energy and, therefore, of a single wavelength (10.6 μm).

After relaxing from the [0 0 1] state to the [1 0 0] condition the CO_2 molecule must lose a substantial amount of energy to return to its ground state before being re-energised. This de-excitation process generally takes place in two stages; [1 0 0]–[0 1 0] followed by [0 1 0]–[0 0 0] (ground state).

The [1 0 0]–[0 1 0] de-excitation is most commonly achieved by a resonant vibrational energy transfer between a CO_2 molecule in the [1 0 0] energy state and another which is in the ground state. The [1 0 0] energy level is of approximately twice the energy of the [0 1 0] state, and therefore a collision between a CO_2 [1 0 0] molecule and a CO_2 [0 0 0] molecule will redistribute the energy between them to put them both in the CO_2 [0 1 0] state:

$$CO_2\ [1\ 0\ 0] + CO_2\ [0\ 0\ 0] = 2CO_2\ [0\ 1\ 0]$$

This resonant relaxation mechanism is quite efficient, but the subsequent de-excitation of the large number of CO_2 [0 1 0] molecules in a pure CO_2 laser is non-resonant and can cause a "bottleneck" in the energy cycle. The energy of a CO_2 [0 1 0] molecule must be converted to kinetic energy by collision with other atoms or the walls of the laser cavity which contains the gas. This rather slow process reduces the efficiency of the laser, and is the reason why helium is added to the CO_2 in modern high power machines (see subsequent section on the role of helium).

On the subject of resonant energy transfer, it should be noted here that the various energy levels of the [0 0 V_3] ladder ([0 0 1], [0 0 2], [0 0 3] etc.) are almost equally spaced. If a high speed electron manages, for example, to excite a CO_2 molecule to its [0 0 3] condition, this energy can be readily redistributed between the excited molecule and two ground state partners to result in three [0 0 1] upper lasing level molecules which can then contribute to the laser output.

Although the high speed electrons travelling through the CO_2 tend to excite the molecules preferentially to the various levels of the [0 0 V_3] ladder, they also energise the molecules to all the other possible states which do not contribute to the lasing process. In order to improve the performance of the CO_2 laser it was found to be necessary to add two other gases to the lasing chamber: nitrogen and helium. The usefulness of these two gases will be discussed in the next sections.

9.2.3 The Role of Nitrogen

Nitrogen is a diatomic (two atom) molecule and therefore has only one possible vibrational mode. For this reason it is possible to use only one vibrational quantum number (V) to describe its excitation state. Figure 9.3 demonstrates that the nitrogen ($V = 1$) state is very close to the CO_2 [0 0 1] energy level. Collision between a nitrogen ($V = 1$) molecule and a CO_2

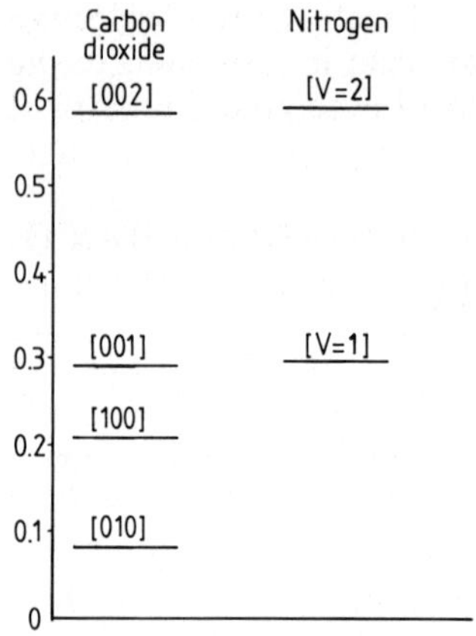

Fig. 9.3. The energy spectra of CO_2 and nitrogen. Note the similarity of the CO_2 [0 0 1] state and the nitrogen [V = 1] state.

[0 0 0] molecule would result in the transfer of energy from the nitrogen to the CO_2 to result in nitrogen (V = O) and CO_2 [0 0 1]. In an electric discharge, nitrogen molecules are efficiently energised to their $V = 1$ level and subsequently pass on this energy to neighbouring CO_2 molecules by collision. In this way the nitrogen can be used to transform the wide range of energies available from the electric discharge into "packages" of energy capable of exciting the CO_2 up to its upper lasing level.

If a nitrogen molecule becomes excited beyond its $V = 1$ condition the energy can still be utilised by the lasing process. The energy levels of nitrogen are almost equally spaced and are all close to the values of the equally spaced CO_2 [0 0 V_3] levels. A collision between a nitrogen ($V = n$) molecule and a CO_2 [0 0 0] molecule will result in an energy transfer, and give a nitrogen ($V = n - m$) molecule and a CO_2 [0 0 m] molecule. It was shown in the previous section that a CO_2 [0 0 m] molecule (where $m > 1$) can subsequently give rise to a number (m) of CO_2 [0 0 1] molecules which can then participate in the lasing process.

9.2.4 The Role of Helium

It was mentioned earlier that although the de-excitation of CO_2 from the lower lasing level to the [0 1 0] state is efficient, the final relaxation from the [0 1 0] to the [0 0 0] ground state is slow in pure CO_2. The de-excitation of CO_2 from the [0 1 0] state takes place as a result of collisions with other particles when the [0 1 0] energy is transformed to kinetic energy. The rate at which this collisional de-excitation takes place depends on the nature of the particles. Each CO_2 molecule undergoes approximately 100 de-exciting collisions per second at a pressure of 1 torr, whereas helium atoms at the same pressure experience approximately 4000.

If helium is mixed with the CO_2 the number of collisions the CO_2 [0 1 0] molecules undergo per second is greatly increased and the rate at which they are de-excited down to their ground state is accelerated. In this way the "bottleneck" in the energy cycle of the CO_2 molecule is removed and the efficiency of the laser as a whole is improved.

Other gases with high collision rates such as oxygen, hydrogen and water vapour (100 000 collisions per second at 1 torr) have been found to give an increase in output power but helium is preferred to the others as it aids the lasing mechanism in two ways:

1. By increasing the de-excitation rate of CO_2 [0 1 0] to CO_2 [0 0 0]
2. By increasing the rate at CO_2 [0 0 0] is excited to the [0 0 1] level

Oxygen, hydrogen and water only increase the de-excitation rate, but helium also improves the excitation rate. Helium achieves this improvement by allowing the electric discharge to have a higher electron density and by altering the distribution of energy in the electron stream to favour CO_2 [0 0 1] and nitrogen [V] excitation.

9.2.5 The Energy Cycle of Mixed Gas CO_2 Lasers

The basic energy input to a CO_2 laser is generally a stream of high velocity electrons in the form of a glow discharge. Other mechanisms of excitation do exist, for example, radio frequency and gas dynamic energy sources. These alternative sources of input energy will be discussed in the later section on laser design, but in the interest of clarity the following description will deal only with a mixed gas CO_2 laser excited by a direct current glow discharge. This discussion will also confine itself to the case of continuous wave (c.w.) laser output. Pulsed output will be dealt with later in the laser design section.

The energy output from the laser cavity is divided into two:

1. The 10.6 μm wavelength light emitted from the output window and subsequently put to use.
2. Surplus heat from various energy exchanges which do not contribute directly to the output beam. This heat is removed from the laser cavity by conduction to the cavity walls or passage of the gas mixture through heat exchangers (see following notes on laser design).

Other, more minor, energy losses such as absorption of the laser light and energy consumed in degrading the chemistry of the gas mixture (e.g. $2CO_2 \rightarrow 2CO + O_2$) will be ignored.

Taking the energy cycle of a "typical" CO_2 molecule in the cavity, it continuously undergoes the following interactions.

Excitation. A CO_2 molecule can be excited from its [0 0 0] to its [0 0 1] condition in any of the following ways:

1. Excitation after collision with an electron in the glow discharge.
2. Excitation after collision with CO_2 molecule in the [0 0 n] ([0 0 2], [0 0 3], etc.) state. Naturally, if the collision is with a [0 0 2] molecule, two upper lasing level [0 0 1] molecules result.
3. Excitation after collision with nitrogen ($V = 1$) molecule.
4. Excitation after collision with nitrogen ($V = n$) molecule.

NB: Excitation mechanisms 1, 2 and 4 can, if the energetic levels are too high, transform a CO_2 [0 0 0] molecule to a CO_2 [0 0 n] state from which it must lose quanta of energy by collision before it reaches the required [0 0 1] condition.

The Lasing Reaction. During the first few microseconds of exposure of the gas to the high speed electrons in the glow discharge, photons are generated as a result of spontaneous relaxation of the energised CO_2 [0 0 1] molecules. These spontaneous emissions generate photons travelling in random directions through the gas mixture. If such a photon interacts with another CO_2 [0 0 1] molecule it will stimulate it to relax down to its [1 0 0] state, thereby emitting a photon of the same energy. These two photons then travel in the same direction repeating this stimulated emission process until an "avalanche" of photons has been established. The photons within these avalanches are all direct twins of the first, spontaneously emitted, photon as they are all of the same wavelength (energy), all travelling in the same direction and all in phase. For this reason they can be described as spatially and temporally coherent.

The initial spontaneous emissions generate avalanches of photons in random directions which are usually lost by absorption by the laser cavity walls. This random process is soon rendered insignificant as a dominant avalanche is established along the axis of the laser cavity. The photons which make up this avalanche are reflected backwards and forwards off the mirrors at each end of the cavity and quickly become responsible for the great majority of photon–CO_2 [0 0 1] lasing interactions. Approximately half the light generated by this process is transmitted through the (partially reflective) output window which forms one of the end mirrors of the cavity. The other half of the photons are reflected back into the laser cavity to continue the stimulated emission avalanche and to keep the laser output constant.

De-excitation. Once a CO_2 [0 0 1] molecule has released its photon and relaxed to the CO_2 [1 0 0] state it is important that it is de-excited to its [0 0 0] ground state as quickly as possible. A CO_2 [1 0 0] molecule needs to return to its [0 0 0] state before it can be efficiently excited to the [0 0 1] condition and subsequently contribute another photon to the laser beam.

The de-excitation takes place in two steps:

1. The CO_2 [1 0 0] molecule collides with a CO_2 [0 0 0] molecule. This collision re-distributes the energy between the two to leave them both in the [0 1 0] state.
2. The CO_2 [0 1 0] molecules lose their energy by collision with helium atoms. The vibrational energy of the molecules is converted to kinetic energy and leaves the CO_2 in its ground state.

NB: This transfer of vibrational to kinetic energy can take place by collision with any foreign body (nitrogen molecules, laser cavity walls etc.), but collision with helium atoms is the most likely.

Once a CO_2 molecule has returned to its ground state it is immediately re-excited by the mechanisms reviewed earlier. This excitation–de-excitation cycle continues as long as the gas in the laser cavity is energised by the electric discharge.

9.2.6 Population Inversion

Population inversion is a phrase often used in any discussion on laser physics to describe the energised state which the lasing medium needs to be in before lasing can take place. The population referred to is the number of possible lasing particles (in this case CO_2 molecules) and the inversion is the reversal of the usual thermodynamic condition that at any given temperature most of the particles would occupy the lowest energy state possible.

The distribution of molecules amongst the various possible energy states is usually governed by the laws of thermodynamics as follows. If the difference between an upper energy level and a lower energy level is ΔE then the ratio of the population of molecules in the two states is

$$\frac{\text{Population (upper level)}}{\text{Population (lower level)}} = \mathrm{e}^{-\Delta E/kT}$$

where k is Boltzmann's constant and T the absolute temperature.

Under normal conditions the number of molecules in the lower state will therefore be much higher than the number in the upper level. If this is the case then any photons travelling through the gas which have an energy of ΔE are likely to collide with lower level gas molecules and become absorbed, thereby raising the molecules to the upper energy level.

The lasing action is quite the opposite to absorption as it involves the multiplication of the original number of photons by simulated emission. For this to be possible the lasing medium must have more molecules in the upper energy state than the lower. This means the usual population ratio has been inverted, hence the phrase "population inversion". This condition is possible in the highly energetic environment of the electric discharge in the laser cavity.

Another piece of jargon which would be best explained at this point is "pumping". Pumping is the word used to describe the excitation which results in the population inversion. For example, if the energy input to a CO_2 laser cavity is electrical, the laser is electrically pumped; some types of laser (e.g. Nd:YAG – see Chap. 8) are energised by high power flashlamps and are therefore "optically pumped" etc.

9.2.7 Laser Modes and TEM Nomenclature

Within the laser cavity the geometry of the end mirrors and their position with respect to each other define a resonator. In all resonators there exist resonances or normal modes. These modes are the result of self sustaining oscillations which become the dominant feature of the energy storage in the laser cavity. The modes can be described by using a three-dimensional index TEM_{mnq} (where TEM stands for Transverse Electromagnetic Mode) where m, n and q define the number of nodes in the energy distribution in the X, Y and Z directions (nodes are zones of low energy). The Z direction is generally taken as the line joining the two end mirrors and the number of nodes in this direction is equal to the number of half wavelengths of light between these mirrors. The number q is therefore generally very large and is typically of the order of millions. On the other hand the number of nodes in the X and Y directions (i.e. across the cross section of the lasing axis) is usually very small, of the order of 0, 1 or 2.

For the industrial user of CO_2 lasers the number of half wavelengths between the mirrors (q) is of little interest, but the cross sectional energy distribution (m, n) is very important as it is the same as the distribution in the output beam. Thus, it is common for the industry to refer to the mode of the beam by defining its TEM_{mn} mode number.

The numbers m and n define the number of nodes in the cross sectional energy density. To illustrate this point a number of examples are given in Fig. 9.4. Using rectangular co-ordinates the TEM_{mn} number is found by counting the nodes or energy gaps first in the X direction (m), then in the Y (n). The high energy zones have energy and gradients within them which usually consist of a general increase of energy towards the centre of any zone. This feature is demonstrated by the energy cross section given in Fig. 9.4. Two special cases are of the greatest interest to CO_2 laser users; the TEM_{00} mode and the TEM_{01^*}. It will be noticed that these are two of the few axially symmetric pure modes possible, axial symmetry being of prime importance for cutting applications (see Chap. 2).

The TEM_{00} mode has a Gaussian energy distribution, and has the best focusing characteristics of all modes. The highest intensity focused spots are possible from such beams therefore cutting speeds can be maximised and kerf widths minimised. This type of mode is normally generated by a stable cavity type laser (see Fig. 1.19).

The TEM_{01^*} mode, also referred to as the "doughnut" mode, has an energy cross section which is a hollow ring, as shown in Fig. 9.4. The TEM_{01^*} label implies that this mode is a special case of the TEM_{01} mode. In cases where the type of laser cavity shown in Fig. 1.19 produces a TEM_{01^*} mode, the mode is generated as a result of the laser oscillating at high frequency between the two modes TEM_{10} and TEM_{01}. The overlap of these two modes results in the characteristic ring of energy with a low or zero energy centre. On the other hand, it is also possible to generate a doughnut type mode from a completely different design of cavity shown in Fig. 1.18. In this case the light which escapes from the cavity by missing the smaller mirror constitutes the laser output and will, naturally, have a hollow ring configuration. This type of design is typical of very high power ($\geqslant$10 kW) CO_2 lasers as it allows a large volume of gas to be excited. The name TEM_{01^*} is convenient to describe this output beam, but readers with an interest in laser physics must bear in mind that in this case the TEM_{nm} label does not describe the mode of energy resonance in the laser cavity. It was mentioned in Chap. 1 that doughnut modes of this type are more suitable to welding and heat treatment than cutting, as a result of their poor focusing characteristics.

Although most commercially available high power lasers are designed to produce TEM_{00} Gaussian or TEM_{01^*} doughnut output modes, certain designs generate a more complex mode which is a combination of one or more modes. Such modes include a Gaussian type with a small depression in

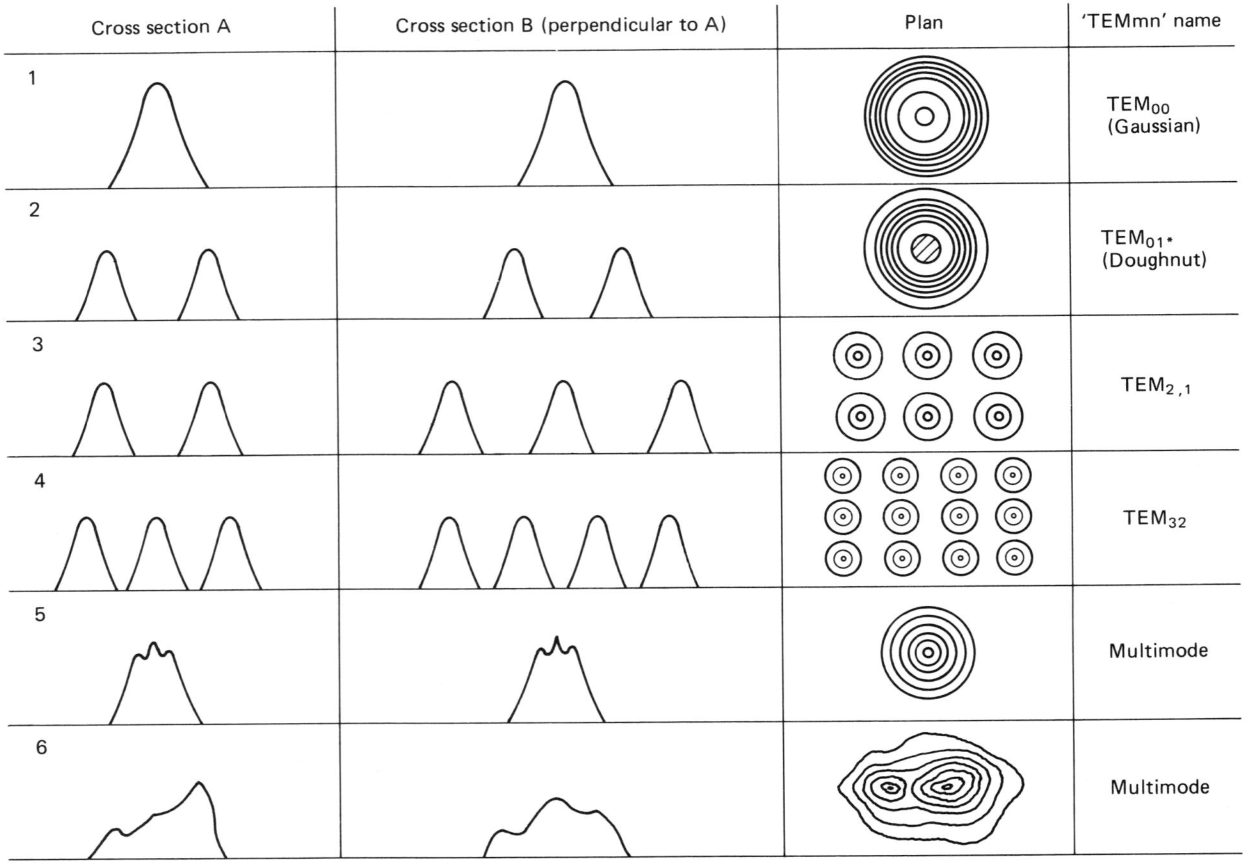

Fig. 9.4. A selection of modes and their TEM_{mn} descriptions. To work out the *mn* numbers simply count the number of energy gaps in the mode in the left to right direction (*m*) and repeat the process in the vertical direction (*n*). Two complex profile multimodes are also shown here to demonstrate the difficulty in accurately naming such modes by the TEM_{mn} system. The axially symmetric multimode (5) would probably cut almost as well as a TEM_{00} mode of the same power. On the other hand, the asymmetric multimode would probably not be suitable for cutting.

the centre or a ring of low energy at a radius between the centre and the edge of the beam. These modes are difficult to accurately name by the TEM system and are often referred to as "multimode". Modes of this type do not focus as well as a Gaussian TEM_{00} type beam, but can still cut with excellent speeds and quality as long as the mode is axially symmetric. Multimodes are generally the result of a deliberate optical design feature which maximises the power output from a certain volume of gas by increasing the level of lasing in the gas some distance from the central lasing axis. However, the term multimode is not fully descriptive and might imply a disorganised energy distribution which would not be suitable for cutting. The multimode shown in Fig. 9.4(b) is of this type.

9.3 Aspects of the Design of CO_2 Lasers

9.3.1 Introduction

The following sections will describe various features of the types of CO_2 laser used for materials processing. These machines range from output powers if a few watts which are only suitable for cutting a limited number of non metals, to large installations generating over 10 kW which are not suitable for cutting but can be used very effectively for welding and surface treatments. In the middle of this range lie the lasers which are most commonly used for cutting. As a general rule CO_2 lasers which are sold for industrial cutting purposes have the following characteristics:

1. *Output power:* 5–3000 W (most commonly ~1000 W or ~2000 W). (Lasers with powers of less than 200 W are not generally suited to metals cutting.)
2. *Method of excitation*: direct current glow discharge or radio frequency excitation. Excitation (and therefore output) can be continuous wave (c.w.) or pulsed.
3. *Optical cavity*: stable cavity (long thin tube type) often folded with mirrors at each bend.
4. *Laser gas cooling*: slow flow (conduction to cooled laser cavity walls) or fast flow (rapid gas circulation through heat exchangers).
5. *Output mode*: always axisymmetric. Often TEM_{00} Gaussian, but other modes have been used to good effect. The output beam diameter is usually less than 25 mm (1.0 in) and is commonly between 12 and 18 mm (0.5 to 0.7 in).

It is worth bearing in mind that CO_2 laser cutting machines are generally members of the subgroup which satisfy the above constraints. The following notes will cover CO_2 lasers in more general terms, a subject which conveniently divides up into three sections:

Methods of exciting the lasing gas mixture
Methods of cooling the lasing gas mixture
Designs of optical cavity

9.3.2 Methods of Exciting the Lasing Gas Mixture

There are many ways of exciting gas molecules to generate laser radiation and, in the search for higher powers, most of these methods have been investigated.

High power CO_2 laser outputs have been achieved using energy input sources as diverse as electron beams and gas dynamics. In the case of a gas dynamic laser a heated gas is compressed and then rapidly expanded by propulsion through an array of nozzles into the lasing chamber. The lasing action takes place as a result of the population inversion possible in a gas after it has experienced very rapid expansion from a compressed high temperature state.

The two types of excitation which dominate the CO_2 laser market for materials processing are direct current (DC) electricity in the form of a glow discharge and the more recent development of radio frequency (RF) excitation. Both methods involve the bombardment of the gas mixture with electrons which have been accelerated by a high voltage electric field. Advantages claimed for RF devices include improved power control and the fact that the electrodes can be mounted outside the laser cavity, a feature which minimises contamination of the gas mixture with metal vapour. These advantages have to be weighed, however, against the disadvantages of the need for RF shielding for safety reasons and the appreciable increase in capital cost of these machines, compared with DC excited lasers.

Most CO_2 lasers are equipped to produce either a continuous wave (c.w.) or pulsed output. The pulses are generated by periodically interrupting the energy input to the laser cavity. It is possible to generate pulses with peak powers which are higher than the equivalent c.w. power by a factor of two or more. A pulsed output can have advantages to the cutting process as it allows a precise, variable control of the output power of the laser. For example, the pulse length, separation and frequency can all be CNC controlled during cutting to rapidly change the output power of the laser if a workpiece has a sudden change in cross sectional thickness. High peak power pulses can also improve cutting performance when profiling high reflectivity materials such as aluminium.

9.3.3 Methods of Cooling the Lasing Gas Mixture

Carbon dioxide lasers have efficiencies of between 10% and 20% and therefore generate a great deal of surplus heat. The lasing mechanism described in previous sections takes place only if the gas mixture as a whole is not allowed to overheat.

Cooling of the gas mixture is, therefore, a major concern when designing a laser and one of two systems of cooling are usually employed:

1. *Slow flow cooling*. Shown schematically in Fig. 1.11, this method relies on direct cooling of the lasing gas by conduction to the walls of the lasing cavity. The cavity walls are themselves cooled by a coaxial jacket of cooled, flowing water, oil or other liquid.

2. *Fast flow cooling*. This technique is shown schematically in Fig. 1.12 and involves the rapid circulation of gas from the laser cavity, through heat exchangers and then back into the cavity.

The two systems have advantages and disadvantages as follows:

Slow Flow. Advantages: extremely simple design with a minimum of moving parts. Vacuum and coolant pumps can easily be replaced by local suppliers. Minimum vibration and noise. Generally superior TEM_{00} Gaussian modes compared with fast flow, as a result of minimum gas turbulence, vibration etc.

Disadvantages: the cooling process is rather inefficient and therefore the power output per unit length of laser cavity is limited to ~50 W/m. This means that the lasers themselves are larger than fast flow machines of the same power, but this can be offset against the need for a large heat exchanger needed by the fast flow machines. For reasons of size, slow flow machines are usually restricted to powers of up to 1200 W.

Fast Flow. Advantages: the increased efficiency of the heat extraction system means that high power lasers can be made smaller. Output powers of more than 250 W per metre of cavity length are possible. Moderate sized multi-kilowatt machines can be made using this technology, although cutting machines tend to be limited to powers of 2 kW or less. In applications where space is essential, for example where the laser needs to fit into a production line, smaller lasers can be a great advantage.

Disadvantages: this type of laser encorporates a fast flowing vacuum sealed gas transport system. These systems often encorporate devices such as Roots blowers which are not as freely available as vacuum and coolant pumps. Failure of such specialist devices can lead to prolonged downtime and large maintenance costs. The machines as a whole are noisier than their slow flow counterparts and are more prone to vibration linked problems. Modes are generally inferior to those of slow flow machines although this may be outweighed by the increased power available. As a general rule, maintenance of these machines is needed more frequently, and repairs are more difficult than for the less complex slow flow machines. Modern designs are improving the reliability of these machines, and this trend is helping to reduce maintenance problems to a minimum.

9.3.4 Designs of Optical Cavity

The simplest design for the optical cavity of a CO_2 laser would be two flat mirrors facing each other across a volume of excited gas. This arrangement is called a Fabry–Perot resonator, and is not generally used for the design of high power CO_2 lasers because it suffers from high diffraction losses. The design of CO_2 lasers has been dominated by the use of mirrors with spherical curvature, usually in combination with flat mirrors. Stable cavity devices (see Fig. 1.19, page 21) are so called because a photon could oscillate backwards and forwards between the two end mirrors of the system indefinitely if it is repeatedly part of the reflected portion of the avalanche. On the other hand, the potential lifetime of any photon in an unstable cavity (see Fig. 1.18, page 19) is very short as the system is designed to expel any photon after only a low number of passes between the mirrors.

Many of the long thin tube stable cavity devices have the correct mirror curvature to generate a TEM_{00} Gaussian beam. This arrangement does not, however, involve the maximum volume of gas in the lasing reaction. In an attempt to increase output powers per unit length, some laser manufacturers have designed the curvatures of the internal mirrors to produce a mode of the type shown in Fig. 9.4(5). Although it is possible to raise the output power in this way the beam cannot focus to as small a spot as the Gaussian distribution. It has been claimed that although the kerf width is increased as a result of the larger focused spot size, the flow dynamics in this broader cut zone are improved for certain metals.

In general, however, the laser cutting industry has concentrated development in favour of Gaussian distribution modes. The minimum diameter, maximum intensity "hot spot" which can be generated by focusing such a beam allows cutting to take place by producing the maximum amount of thermal damage over the narrowest area.

Chapter 10

Some Aspects of the Physics and Chemistry of Laser Cutting

(Based on work carried out in collaboration with Dr. A. Ivarson of the University of Lulea, Sweden.)

10.1 Introduction

Although a great of scientific and commercial research has been carried out on the subject of laser cutting, certain aspects of the process are not yet completely understood and researchers in the field are still kept extremely active. The work which has been carried out so far has led to the rapid growth of the industry and from a practical point of view a deeper understanding may have only a small effect on future industrial development. On the other hand, theoretical and experimental research has often given rise to enormous commercial rewards in industries which were thought to be in a stable condition.

The laser cutting process involves physical and chemical dynamics and can only be fully understood by reference to several different disciplines, including optics, heat and mass transfer, fluid dynamics, phase equilibria and chemical thermodynamics.

Bearing in mind the complex nature of the interplay between these subjects, it is not surprising that our understanding of the process is incomplete. The remainder of this chapter will present a few discussions on various aspects of the cutting process under the following headings:

1. The energy balance in the cut zone and its relationship to the efficiency of the process.
2. The role of oxidation when cutting steels.
3. Conductive losses experienced when cutting steels.
4. Notes on reflected, transmitted, radiated and convective losses from the cut zone.
5. Notes on the focusing characteristics of CO_2 laser beams.

10.2 The Energy Balance in the Cut Zone and Its Relationship to the Efficiency of the Process

During laser cutting a dynamic equilibrium exists in the cut zone which balances the "incoming" energy and material with the "outgoing" energy and material.

Inputs to the cut zone are usually only

1. The laser energy
2. The pressurised cutting gas jet (which may or may not be chemically reactive with the workpiece)

Outputs from the cut zone are more numerous and complex in nature:

1. Solid, liquid or gaseous material created in the cut zone and ejected to produce the cut
2. The exhaust of the pressurised cutting gas jet (which may or may not have undergone chemical reaction with the workpiece, e.g. $yO + xFe = Fe_xO_y$ in the case of mild steel cutting)
3. Energy in the following forms:
   (a) Conducted heat
   (b) Reflected laser light
   (c) Radiated light
   (d) Convected heat
   (e) Transmitted laser light (where "transmitted" does not necessarily imply that the light has passed through the material being cut – a small proportion of light will pass directly through the cut zone without interacting with the cut front at all)

A simple energy balance for laser cutting can be expressed as

Energy supplied to the cut zone
= Energy used in generating a cut + Energy losses from the cut zone (by conduction, radiation etc.)

The word "losses" is used here to describe energy which does not contribute to the removal of material from the cut zone.

It is the aim of this chapter to analyse various aspects of this energy balance, paying particular attention to thermal losses from the cut zone. It can be shown by a simple theoretical analysis and experimental programme (Sects 10.4 and 10.5) that the proportion of energy lost from the cut zone decreases with increasing cutting speed. This means that the efficiency of the cutting process increases with increasing cutting speed, a phenomenon which has a profound effect on cutting speeds and on the maximum thickness any material which can be cut by a particular laser.

The earlier statement:

Energy supplied to the cut zone
= Energy used in generating a cut + Energy losses from the cut zone

can be expressed by the following formula:

$$(P - b)t\left[\frac{100 - r_f}{100}\right] = (E_{cut}ldk) + [(t\pi\, dk/2)(A + B + C)] \quad (10.1)$$

Considering a laser of power P cutting a line of length l in a time t, where

b = laser power transmitted through the cut zone without interaction with the cut front

r_f = the reflectivity of the cut zone expressed as a percentage (NB: $100 - r_f$ = absorptivity as a percentage)

E_{cut} = specific energy needed to melt and remove one unit volume of material from the cut zone

d = material thickness

k = kerf width

A = conductive loss function
B = radiative loss function
C = convective loss function
} power per unit area

It will be noticed that the reflective and transmitted losses from the cut zone are dealt with in the left hand side of the equation, whereas the conductive convective and radiative losses are included in the right hand side.

The logic behind this approach is the consideration that the losses from the cut zone can be divided into two types; primary and secondary.

Primary losses are those which leave the cut zone as they entered, as 10.6 μm radiation. These include only transmitted and reflected laser light.

Secondary losses are those which leave the cut zone after thermal transformation of some kind. These include the conducted, convected and radiated losses. (The broad spectrum of radiated light may include a small proportion of 10.6 μm light, but this did not originally emanate from the laser and is therefore part of the secondary losses.)

The secondary losses are a function of the temperature of the cutting front and its surface area in contact with the surroundings. The conductive losses are made from the convex face in contact with the substrate and the convective and radiative losses are made from the concave face exposed to the surrounding atmosphere as shown in Fig. 10.1. This schematic takes the simplified view that the cut front is the shape of half a thin walled tube which has been bisected along its length.

Other points and simplified assumptions in Eq. (10.1) which may need some clarification are:

b (transmitted losses). During cutting it is often the case that the trailing edge of the cut front does not extend to the full diameter of the incident laser beam. A proportion of the available light therefore passes straight through the kerf without interacting with the cut front.

r_f. The reflectivity of the cut zone will be much less than the figures quoted for solid materials at ambient temperature. The cut zone has a higher absorptivity as a result of its high temperature, the presence of absorptive oxides, its shallow angle of incidence to the laser beam, its roughness and its absorptive layer of vapour.

The area surrounding the cut zone reaches the temperature at which it can ignite in the oxygen stream and a widespread reaction takes place.

The second effect is due to the low average temperature of the melt, compared with that created in thinner sections. This reduction in temperature is a result of the fact that so much energy is now being "wasted" by conduction, convection and radiation that not enough is available for the cutting process. Other features of the process such as chemical dynamics and fluid flow are also important contributors to the breakdown of the process, but can be ignored for the purpose of this argument. (High pressure inert gas laser cutting of metals also experiences a rapid degradation in cutting mechanism above a certain thickness for a particular laser–material combination.)

Low conductivity materials such as acrylic can be cut at much greater thicknesses as their cutting efficiency is a great deal higher than that of metals (ignoring the effects of oxidising reactions). An increase in thickness of the material does lead to a build up in the proportion of energy "wasted" but this is a much more gradual process than in the case of metals. The relatively low temperature of the cutting front keeps radiative heat losses to a minimum. (Radiative heat loss is proportional to T^4, where T is the absolute temperature.) Conductive losses are low as the thermal conductivity is small. For these reasons, many polymers can be cut at thicknesses an order of magnitude larger than those for metals.

In certain cases the losses from the "cut" zone are so large that cutting cannot be initiated. The most common example of this is copper sheet which cannot be cut at all below a critical power density. For example a c.w. laser of below 500 W will be unable to initiate a cut zone in copper at thicknesses above foil grades. In this case the primary heat loss is due to reflectivity, aided by the high thermal conductivity.

10.2.3 The Implications of the Energy Balance Argument to Changes in Laser Power

Equation (10.1) implies that the cutting efficiency is proportional to the cutting speed and therefore to the laser power P. In practice this holds true only in a few specific cases and it is just as likely that the efficiency will remain steady or decrease with laser power if a broad range of applications is investigated. Cutting speeds will generally rise with laser power in an approximately proportional manner, except where very high power machines ($\geq$3 kW) are used. In the case of these higher powered machines, cutting can be considerably slower than expected. The reasons for this uncertainty of the effect of increasing power include the following:

1. If two lasers are being compared (e.g. 500 W and 1200 W) it is probable that the higher power machine will have an inferior mode quality which will not focus to as small a spot as the low powered machine. This larger focal spot will produce wider cuts which will render the process less efficient as more material will have to be removed to generate the cut. This point is particularly relevant to machines with powers in excess of 3 kW.

2. The fluid dynamics of the cut zone play a strong role in determining the material removal (cutting) rate. At the greater cutting rates possible with a higher powered machine the inclination and geometry of the cut front may change and this will affect removal rates. The flow of the cutting gas jet may have to be increased to cope with the increased material removal rate. Above a limiting cutting speed the viscosity of the melt may become the rate determining factor.
3. As a result of the reduction in thermal losses to the workpiece when cutting at higher speeds, the thermal gradients around the cutting zone are more severe. The material along the cut line has not been as effectively preheated by the moving cut front and therefore requires more energy to become melted and ejected.

10.3 The Role of Oxidation when Cutting Steels

10.3.1 Mild Steel

During laser cutting of mild steel the use of oxygen as a cutting gas has two major advantages:

1. The exothermic reaction of iron with the oxygen provides a second energy input to the cut zone. This substantially increases the cutting speeds possible.
2. The oxidation reaction generates an oxidised molten pool which has a low surface tension and a low adhesion to the surrounding solid material. This ensures easy removal of material from the cut zone and a dross free cut edge.

The amount of energy generated by the oxidation reaction in the cut zone depends upon the amount of material oxidised and the type of oxidation reaction. Iron can undergo any of three oxidation reactions to yield either FeO, Fe_2O_3 or Fe_3O_4. Experimental work (Refs 20 and 21, Chap. 11) has shown that it is possible to identify the amount and type of oxidation experienced by the iron by collecting the oxidised material as it leaves the bottom of the cut zone. This material leaves the cut zone as incandescent sparks which are spherical in nature and have diameters of between 50 and 500 μm. These can be collected in a (non-oxidising) liquid bath of water and glycol for subsequent chemical analysis. The results of an experiment of this type are shown in Table 10.1. It is clear that the thickness of the material makes little difference to the chemistry of the ejected material and so average values of free iron, oxidised iron and oxygen can be used to determine the chemical reaction which was common to all the samples.

The particles are obviously made up of a mixture of iron and oxidised iron and this is confirmed by Fig. 10.2, which shows an etched cross section of a typical example. The similarity in the proportions of oxidised and non-oxidised iron indicates that, as a first approximation, half the ejected iron was oxidised during the cutting process. Having established the extent of the oxidation reaction, it is now important to identify the type of reaction, as each of the three possible oxidation reactions generates a different amount of heat.

ference between three sample thicknesses cut at different speeds and the average chemistry was as follows:

Unoxidised iron (wt%)	43.0
Oxidised iron (wt%)	19.5
Unoxidised nickel (wt%)	10.9
Oxidised nickel (wt%)	0.67
Total chromium	15.2
Total oxygen	10.7

(NB: It is not possible to estimate directly the amount of chromium oxide present by wet chemical analysis.)

This chemical analysis combined with microscopic observation of sectioned particles revealed that they consisted of a central metallic core surrounded by a layer of oxide similar to their mild steel counterparts.

A calculation based on the total oxygen content was used to estimate the extent of the chromium oxidation and a general formulation of the chemistry of the ejected material could be established (Table 10.3). The formation of the three oxides in Table 10.3 generates a great deal more heat per mole than the simple oxidation of iron to FeO as we can see from the following:

$$2Fe + \tfrac{3}{2}O_2 \rightarrow Fe_2O_3 \quad \Delta H = -826.72\,\text{kJ/mol at } 2000\,\text{K}$$
$$2Cr + \tfrac{3}{2}O_2 \rightarrow Cr_2O_3 \quad \Delta H = -1163.67\,\text{kJ/mol at } 2000\,\text{K}$$
$$Ni + \tfrac{1}{2}O_2 \rightarrow NiO \quad \Delta H = -248.23\,\text{kJ/mol at } 2000\,\text{K}$$

It is clear that the energies of formation of Fe_2O_3 and Cr_2O_3 are much larger than the value for FeO of −257.58 kJ/mol.

As an approximation it can be taken that 30% of the iron and chromium which leaves the cut zone is oxidised. (Thermodynamically the NiO formation is negligible.)

From this information we can estimate the power of the oxidation reaction as we did in the case of mild steel (Table 10.4). The output power

Table 10.3. Chemical composition of ejected material

	Fe	Fe_2O_3	Cr	Cr_2O_3	Ni	NiO
Wt%	43.0	27.8	10.6	6.8	10.9	0.87
Mol%	55.4	12.5	14.7	3.2	13.4	0.86

Table 10.4. Oxidation heat evolved during stainless steel cutting

Sample thickness mm	Cutting speed mm/s	Kerf width mm	Material removal rate mm^3/s	Oxidation heat evolved W
1	133	0.27	35.9	578
2	65	0.29	37.7	607
3	42	0.32	40.3	649

Average power of the oxidation reactions: 611 Ws

of the laser used for this experiment was 900 W, and so in this case the oxidation reactions supply approximately 40% of the energy to the cut zone. This proportion can be expected to change with gas pressure, and may be substantially larger when thicker sections (⩾6 mm) are cut at supply pressures of 10 bar or more.

10.4 Conductive Losses Experienced when Cutting Steels

Section 10.2 presented an energy balance argument which made it clear that energy being fed to the cut zone was either utilised during the cutting process or was dissipated by conduction, transmission, reflection, convection or radiation. Section 10.3 has demonstrated the importance of the oxidation reactions as a thermal input to the cut zone when cutting steels. Equation (10.1) for steel cutting should therefore be re-written as

$$(P - b)t\left[\frac{100 - r_{\mathrm{f}}}{100}\right] + H = (E_{\mathrm{cut}}ldk) + [(t\pi dk/2)(A + B + C)] \tag{10.5}$$

where H is the power of the oxidation reaction calculated as in Sect. 10.3 and expressed in watts.

Having investigated H in the previous section, it is now important to analyse the various ways in which energy is dissipated from the cut zone. This section will deal with the major source of heat loss for metals: conduction. Section 10.5 will include more general notes on transmission, reflection, convection and radiation.

It is of course a gross simplification to discuss heat conducted away from the cut zone as wasted energy as far as the cutting process is concerned. It is by this conduction of heat that the area in front of the cut zone is given energy which will eventually result in melting and/or oxidation. In an ideal situation, however, the most effective use of the energy input to the cut zone would be if only material which left that zone experienced heating. This would result in maximum cutting efficiency and an unheated cut component. As the thermal conductivity of the material being cut increases this ideal situation becomes more and more remote until in some cases (e.g. thick section copper) cutting becomes impossible. For the more usual case of the cutting of steels etc., it can be said that heat which has left the cut zone to generally warm the eventual cut component has not been directly involved in the cutting process. The magnitude of this conducted heating is therefore an indication of the level of inefficiency of the cutting processes and can be described as "wasted" energy or conductive "losses" from the cut zone.

A simple experiment can be carried out to determine the conductive losses from the cutting process. If the cut zone moves in a straight line it loses heat equally to either side of that line. If the line is curved into a circle or steered into any other enclosed shape the "component" cut out of the sheet will be warmed by half the total energy lost by conduction during the cutting process. If the heat absorbed by the component is measured and its perimeter and the cutting speed are known, it is a simple matter to establish the conductive loss per second from the cut zone.

To investigate conductive losses during steel cutting, a number of 50 mm diameter discs were cut from mild and stainless steel of various thicknesses. Immediately after cutting the discs were immersed in water and the heat absorbed by them by conduction was measured by calorimetry (Ref. 31, Chap. 11). The results of this experiment are given in Table 10.5.

The calculation to determine the conductive losses in watts is carried out from the experimental data as follows. Taking sample A,

Cutting speed: 6.4 m/min
Heat gained by disc during cutting: 544 J

Therefore total heat lost by conduction from the cut zone during the cutting of the disc: $2 \times 544 = 1088\,\text{J}$.

Time taken to cut the 50 mm diameter disc: 1.47 s.

Therefore

$$\text{Conductive losses} = 739\,\text{J/s} \quad (739\,\text{W}).$$

It is convenient to express this in watts because the losses can thereby be directly compared with the input power from the laser and the oxidation reactions.

It is clear from Table 10.5 that the conducted losses rise rapidly as the thickness of the material is increased for both stainless and mild steels. This was to be expected from the discussions in Sect. 10.1 where it was pointed out that when cutting thicker material the cut front would be larger and would therefore be able to dissipate more heat. As a first approximation the kerf width was constant throughout the samples and had a value of ~0.3 mm. This being the case, and assuming a similar (Fig. 10.1) geometry for the cut fronts, the area of the solid–liquid boundary of the cut zone is directly proportional to the thickness of the specimen. It is across this

Table 10.5. Conductive losses from steel cutting

Sample	Thickness mm	Oxygen pressure bar	Cutting speed m/min	Heat gained per disc J	Conductive losses W
Mild steel					
A	1.55	3.0	6.4	544	739
B	2.9	2.3	3.15	1314	876
C	4.8	1.5	1.8	2919	1112
D	8.0	1.0	0.8	7698	1305
Stainless steel					
E	1.2	6.5	7.7	309	502
F	1.9	6.5	4.08	1077	936
G	3.25	6.5	2.4	2896	1485
H	5.00	8.0	1.53	6521	2103

Laser power 800 W; lens focal length 127 mm; nozzle diameter 1.2 mm; nozzle–material standoff 1.5 mm; cut disc diameter 50 mm.

Table 10.6. Conductive loss per unit thickness

Sample	Total conductive losses W	Thickness mm	Conductive losses per unit thickness W/mm
Mild steel			
A	739	1.55	477
B	876	2.9	302
C	1112	4.8	232
D	1302	8.0	163
Stainless steel			
E	502	1.2	418
F	936	1.9	493
G	1485	3.25	457
H	2103	5.00	420

solid–liquid boundary that heat is being lost by conduction from the cut zone and so it is interesting to divide the number of watts lost by conduction by the thickness of the material (Table 10.6).

Within the limits of experimental error the figures for the stainless steel samples remain fairly constant, which means that the cut front loses approximately the same amount of heat per unit area whatever the material thickness. This is remarkable because the laws of heat transfer make it clear that the rate of conductive energy loss from a hot zone is related to the thermal gradient between the hot zone and its cooler surroundings. In the case of laser cutting, the thermal gradient around the cut zone will become more shallow as the process is slowed down to cut thicker sections. The mild steel samples follow the expected trend in decreasing the watts per unit area lost by conduction from the cut zone as the thermal gradient becomes shallower. The reason behind the more consistent value for stainless steel must lie in the energetic conditions in the cut zone. It is possible that oxidation reaction rates and cut front temperatures are increased in deeper cut zones as the static pressure of the incident oxygen jet is increased together with the laser–material interaction time.

The main point of the information given in Table 10.5 is the indication that thermal losses by conduction from the cut zone are extremely large when cutting steels. The rapid increase in the amount of heat lost as the thickness of the workpiece increases was predicted in Sect. 10.2 and is clearly the reason behind the non-linear reduction in cutting speed with thickness and the eventual collapse of the process at a maximum section.

The same argument holds true for all metals, and the following section will demonstrate that the other types of thermal loss from the cut zone are small by comparison.

10.5 Notes on Reflected, Transmitted, Radiated and Convective Losses from the Cut Zone

These notes follow on directly from the comments made about thermal losses from the cut zone at the beginning of Sect. 10.4. Although the major thermal loss when cutting metals will usually be by conduction, energy is also dissipated from the cut zone as a result of reflection, transmission, radiation and convection. These types of loss can be divided into two groups:

1. Reflection and transmission are "primary" losses (see Sect. 10.2). In other words the energy leaves the cut zone in the same form as it entered, as 10.6 μm wavelength light.
2. Radiation and convective losses are classified as secondary because they are the result of a thermal transformation in the cut zone.

10.5.1 Reflected and Transmitted Losses

It was mentioned in Chap. 3 that the reflectivity of solid metals at ambient temperature is very high to the beam generated by a CO_2 laser. Fortunately for the cutting process the absorptivity rises from a small percentage to a much larger value as the material is heated and a cutting front is established. The high absorptivity of the cut front is the result of a number of factors:

1. The cut front is at a high temperature (absorptivity increases with temperature).
2. The beam strikes the cut front at a shallow angle (absorption is high at shallow angles of incidence).
3. The melt has a rough surface which in many cases is covered with highly absorptive oxides.
4. The melt has a vapour associated with its surface which is often highly absorptive.

All these factors combine to ensure that the absorptivity of the cut zone is high, although the cutting process still involves the interaction of a light beam with a target material and some reflection must be expected.

Transmission of light directly through the cut zone can be clearly observed if any material is cut too slowly for a particular laser power. The light itself is, of course, invisible, but its heating effect on the base of the cutting table or any object below the cut zone can be easily seen. During optimum speed cutting this transmission of surplus energy will be minimised, but a small amount of "leakage" is likely as the trailing edge of the beam often lags the trailing edge of the cut front.

An experiment to investigate the transmission and reflection losses from the cut zone was set up as in Fig. 10.3. The acrylic block is used to collect any reflected or transmitted light which is emitted from the cut zone in the form of a mode print (see Sect. 5.6.2). The main experimental programme involved cutting various thicknesses of acrylic at a constant laser power and cutting speed, while observing how the amount of reflected/transmitted light reduced as the workpiece thickness was increased (Ref. 35, Chap. 11). The cuts were all of the same length and therefore the same duration, an important point as the volume of material evaporated from the acrylic block could be used to compare directly the surplus energy emitted from the bottom of the cut zone. Miyamoto et al. (Ref. 66, Chap. 11) have demonstrated that the volume of acrylic evaporated by a laser beam in a certain

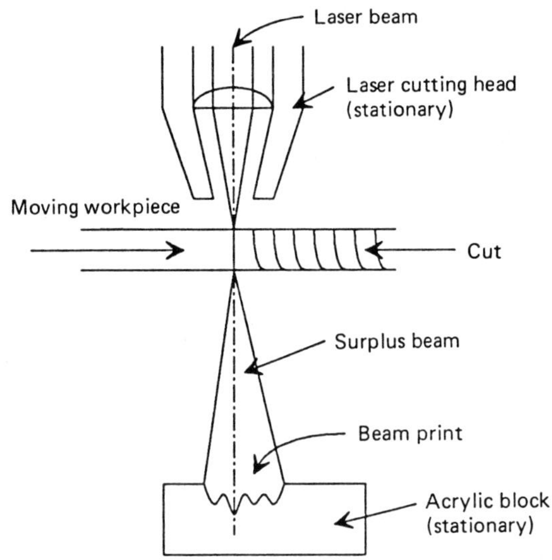

Fig. 10.3. The experimental set up for the investigation into transmission and reflective losses from the cut zone.

time can be used as a measure of laser power. They established a specific ablation energy for acrylic of 3000 J/cm^3, i.e. a 3000 W laser would evaporate 1 cm^3/s. This principle was used to estimate the laser energy used during cutting of the workpiece (by volumetric measurement of the cut) and that consumed in evaporating the acrylic block (by volumetric measurement of the mode print). In this case it was found for all the sample thicknesses cut that the addition of the two estimates gave a constant figure of ~825 mm^3, equivalent to a laser power of 321.4 W which compares very well with the actual laser power used (325 W). Therefore, for cutting acrylic, we can say that any laser energy which is not employed in cutting the workpiece exits from the bottom of the cut zone as reflected or transmitted light. This view is supported by the low thermal conductivity of acrylic (~0.15 W/m/°C), compared with steel (~45.0 W/m/°C, Ref. 80, Chap. 11) and the low cut front temperature (~600 °C, Refs 43 and 44), both of which would minimise conductive and radiative losses from the cut zone.

On a qualitative level the mode prints gave a very interesting insight into the nature of the reflection and absorption in the cut zone. Figure 10.4 shows photographs of the mode prints obtained from below the cut zone when cutting acrylic 15, 18 and 21 mm thick (21 mm was the thickest section which could be cut reliably with this laser power (325 W) at the given speed of 13 mm/s). The cuts in each case were 100 mm long and, as a result of the constant cutting speed, the acrylic mode prints were all exposed to surplus beam for 7.7 s. The evaporated volumes and equivalent laser powers for the three mode prints are given in Table 10.7.

The geometry of the mode prints is very important as they all demonstrate the presence of interference fringes. Fringes of this type are generated when two light beams of identical wavelength arrive in the same target area by slightly different paths. Two beams in this situation give rise to alternating

a

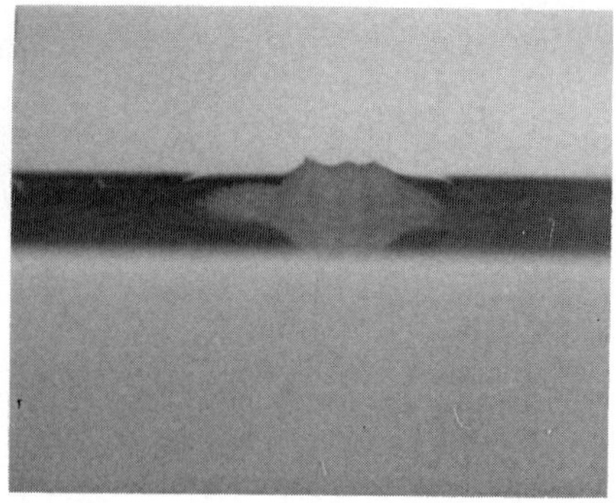

b

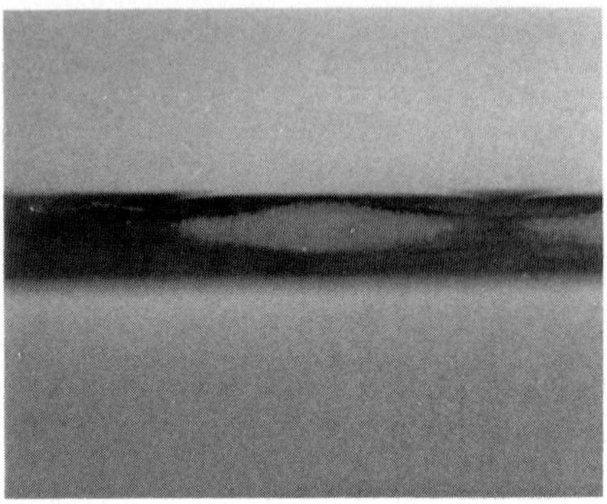

c

Fig. 10.4. Photographs of mode prints made as shown in Fig. 10.3. These prints were taken when cutting acrylic of thicknesses **a** 15 mm, **b** 18 mm and **c** 21 mm under identical conditions (laser power 325 W, cutting speed 13 mm/s, cut length 100 mm; therefore mode print duration 7.7 s).

Table 10.7. Mode print volumes

Sample	Workpiece thickness mm	Mode print evaporated volume cm^3	Equivalent laser power W
A	15	0.21	82.0
B	18	0.14	54.5
C	21	0.03	11.5

Specific ablation energy of acrylic: 3000 J/cm^3 (Ref. 65, Chap. 11); ablation time: 7.7 s.

zones of constructive or destructive interference which can usually be observed as bright and dark lines or fringes. In the case of CO_2 laser light irradiating acrylic however, these bright and dark areas are replaced by areas of high and low evaporation rates hence the ridges in the mode prints of Fig. 10.4. Figure 10.5 demonstrates that the interference of reflected and transmitted light will produce a ridged mode print of this type.

If the mode print for the 15 mm cut sample (Fig. 10.4a) can be taken as consisting of a substantial amount of reflected and transmitted light, then the decrease in mode print volume towards zero does not follow the expected trend. If the reflected energy off a 15 mm cut front was X W and the transmitted energy was Y W, it would be expected that if the workpiece thickness increased the value of X would increase as a result of the increase

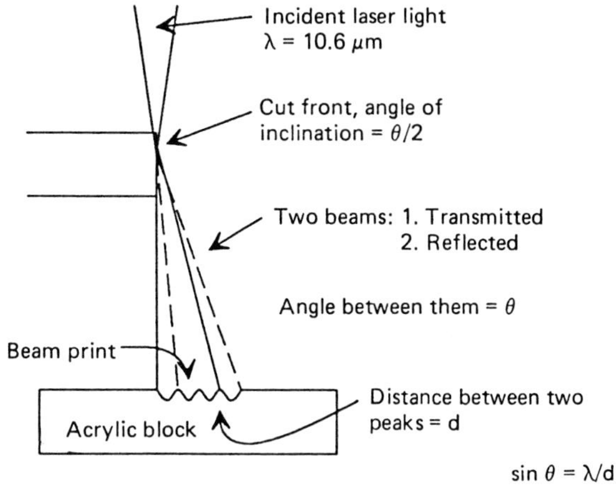

Fig. 10.5. A schematic of the interference of reflected and transmitted beams which generates a ridged mode print.

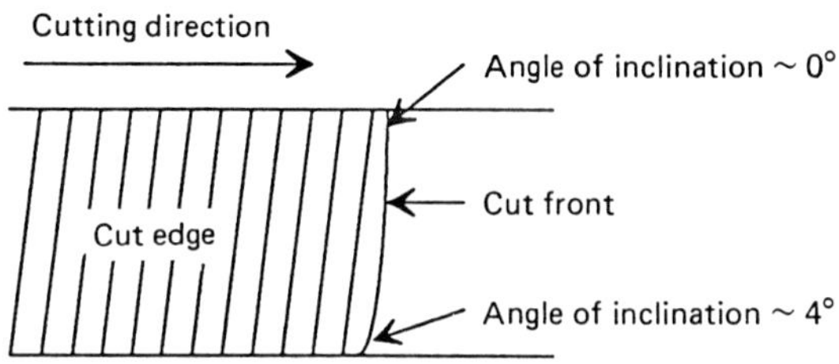

Fig. 10.6. The geometry of the cut front when cutting close to the maximum speed. (At lower speeds the cut front is a straight, almost vertical line.)

in the interaction area between the reflective cut front and the beam. Y would decrease as the lag between the trailing edges of the cut zone and the beam decreased. The reduction of the combined transmitted and reflected losses down to zero indicates a change in the nature of the reflection as the sample thickness is increased. This can be explained by close observation of the change of shape of the cut front as either the cutting speed or the material thickness is increased. When a material such as acrylic is cut either too slowly or with too much power, the cut front is nearly vertical and almost a straight line. In this case part of the beam will pass straight through the cut zone and another part will be reflected off the slightly inclined cut front (Fig. 10.5). These two beams may then combine below the cut front and give a mode print like the one observed for the 15 mm thick sample (Fig. 10.4a). As the cutting speed or the material thickness is increased the cut front geometry changes to the type shown in Fig. 10.6. Here it can be seen that the angle of inclination of the cut front changes from almost vertical at the top to several degrees from vertical at the bottom. Light entering the cut zone from the top may be reflected off the upper part of the cut zone, but this reflection will be at such a shallow angle that the light will re-impinge upon the workpiece further down the cut front. This multiple reflection means that the chances of absorption are high and reflective losses are minimised. The geometry of cut front shown in Fig. 10.6 also minimises transmissive losses because the lag between the trailing edge of the cut zone and that of the beam will rapidly tend to zero.

It should be noted here that this experiment gives much the same results for a wide variety of workpiece materials including other polymers, stainless steel and wood. The important exception is mild steel from which non-ridged mode prints were collected from under the cut zone. This anomaly is probably the result of the rapidly fluctuating geometry of the cut zone (see Chap. 2) which will give rise to a complex dynamic reflection not capable of producing interference patterns.

In summary, it can be said that when the cutting process is optimised (i.e. cutting at the highest reliable speed) the losses from reflection and transmission are a very low percentage of the incident laser energy. This is largely because the curved geometry of the cut front (Fig. 10.6) helps to maximise absorption.

10.5.2 Radiation and Convective Losses

Radiation and convective losses are more difficult to quantify than those covered so far for a number of reasons:

1. The amount of heat involved is generally low (as we shall see in this section).
2. Reliable sampling of the energy is difficult; in the case of radiation losses the geometry of the cut front means that a large proportion of the radiated light will be re-absorbed by the recently cut edges. Convective losses are contained in the gas leaving the bottom of the cut zone which also carries the hot debris ejected during cutting. The gas thus has a dual energy content which is difficult to separate experimentally.
3. The cut front may or may not have a steady state geometry and temperature distribution.
4. Thermal gradients on the cut front surface are inaccessible to direct measurement by optical pyrometry etc., because the front is usually very small and almost completely enclosed by the workpiece material being cut.

In order to show that radiative and convective losses from the cut zone are generally a minor influence on the energy balance a generalised theoretical argument can be used in both cases.

10.5.2.1 Losses by Radiation

The physical laws of radiation make it possible to determine the level of radiation from a hot body if the following details are known:

1. The temperature of the body
2. The surface area of the body
3. The emissivity of the body

The emissivity of a body is a material dependent quality describing its ability to radiate. This has a value from 0.0 to 1.0, the latter being the emissivity of a theoretically perfect absorber and radiator or "black body". For example, soot has an emissivity of more than 0.9, whereas solid steel can vary from ~0.1 (polished) to ~0.8 (oxidised).

The Stefan–Boltzmann law demonstrates that the energy radiated from a hot body is proportional to T^4 where T is the absolute temperature of that body:

$$q = E\sigma T^4 \tag{10.6}$$

where

q = The power radiated per unit area (W/m^2)
E = The emissivity of the material from which the body is made
σ = The Stefan–Boltzmann constant ($5.67 \times 10^{-8}\,W/m^2\,K^4$)
T = The absolute temperature (K)

If we can assume for the sake of argument that the emissivity of the cut front is 1.0 then we can calculate a "worst case" radiative loss if we know the temperature and geometry of the front. If the geometry is taken as being that represented in Fig. 10.1, then it is easy to calculate the maximum radiative output for any cut front at any temperature. Table 10.8 demonstrates this for a typical kerf (0.3 mm wide, 5 mm thick material) and an unusually large one (1.0 mm wide, 10 mm thick).

Optical pyrometry has been used to establish cut front temperatures for a number of materials as shown in Table 10.9. Approximate values relating to thermoplastic and thermoset polymeric materials and steels are indicated in Table 10.8 and it can be seen that in all but the "large kerf–thermoset" case the maximum possible losses by radiation from the cut zone are minor. The actual values of the losses will be smaller than these as the cut front will have an emissivity equal to less than one.

The only case where thermal losses by radiation will have a significant effect on cutting speeds and the efficiency of the cutting process would

Table 10.8. Maximum radiative outputs from the cut zone

Temperature K	°C	Theoretical maximum radiative output W/cm²	Max output from 0.3 mm kerf in 5 mm material W	Max output from 1.0 mm kerf in 10 mm material W	Measured cut front temperatures for various materials
500	227	0.35	0.01	0.05	
1000	727	5.68	0.13	0.89	Thermoplastic polymers
1500	1227	28.8	0.68	4.52	
2000	1727	91.0	2.14	14.29	Steels
2500	2227	220.0	5.17	34.56	
3000	2727	460.0	10.8	72.2	Thermoset polymers and wood
3500	3227	850.0	20.0	133.5	
4000	3727	1400.0	33.0	220.0	
4500	4227	2300.0	54.0	361.3	
5000	4727	3500.0	82.2	550.0	

Table 10.9. Cut front temperatures

Material	Cut front temperature (K)	(°C)
Mild steel	1773	1500
Stainless steel	2123	1850
Thermoplastic polymers	873–1273	600–1000
Thermoset polymers and wood	2273–3273	2000–3000

be when cutting wood or thermoset polymers with a low power. If, for example, die boards are to be cut for the packaging industry, it is common to cut sections thicker than 10 mm with a kerf width of ~1.0 mm. In this case the loss by radiation could exceed 100 W although a large amount of this energy would be re-absorbed by the adjacent cut edges. (This re-absorption does not aid the cutting process but it does reduce glare in the work place.) This loss is substantial if a laser of only ~500 W is being employed. Fortunately the output by radiation is not dependent on the incident laser power and will remain at its ~100 W level whatever laser power is being used. This means that the efficiency of the process will be improved if the laser power is increased. Cutting speeds will, for example, be more than doubled if the laser power is doubled.

10.5.2.2 Losses by Convection

The thermal losses by convection from a hot material to a surrounding gas can be described by a simple formula:

$$Q = Ah(T_f - T_g) \tag{10.7}$$

where

Q = Thermal loss by convection (kW)
A = Surface area of hot zone (m^2)
h = Convection heat transfer coefficient (kW/m^2K)
T_f = Temperature of hot zone (K)
T_g = Temperature of incident gas (K)

Although the formula itself is very simple it involves one extremely complicated factor, h, the convection heat transfer coefficient. To evaluate this number it is necessary to know the properties of the gas (density, thermal conductivity, specific heat etc.) and the type of flow experienced near the hot zone. There are three basic types of convective interaction: free convection, laminar flow forced convection and turbulent flow forced convection. During laser cutting we are generally concerned with the latter of these three types which is, unfortunately, the most difficult to solve by analytical methods. It is outside the scope of this book to attempt such an analysis which, in any case, would have to deal with individual examples of cut zones one at a time. It is, however, a rather simple matter to identify the range of losses which could be expected, because it is well known that for forced convective cooling by a gas the value of h generally lies between 0.01

Table 10.10. Minimum, intermediate and maximum convective losses to be expected from two types of cut zone[a]

Cut front temperature (K)	(°C)	Min. loss (W) ($h = 0.01$)[b] A	B	Int. loss (W) ($h = 0.1$) A	B	Max. loss (W) ($h = 1.0$) A	B
500	227	0.006	0.034	0.06	0.34	0.6	3.4
1000	727	0.02	0.11	0.20	1.1	2.0	11.0
2000	1727	0.05	0.28	0.50	2.8	5.0	28.0
3000	2727	0.08	0.45	0.80	4.5	8.0	45.0
4000	3727	0.10	0.56	1.00	5.6	10.0	56.0

[a] Cut type A: kerf width 0.3 mm, material thickness 6.0 mm; cut type B: kerf width 1.0 mm, material thickness 10.0 mm.
[b] Values of h in kW/m^2K.
Initial cutting gas temperature taken as 20 °C, 293 K.
Cut front geometry assumed to be as in Fig. 10.1.

and $1.0\,kW/m^2K$ (Ref. 82, Chap. 11). This being the case it is easy to calculate the minimum, intermediate ($h = 0.1$) and maximum convective losses for any cut zone as a function of temperature. Table 10.9 is a result of this calculation for two cut types, one with a typical kerf width of 0.3 mm and a material thickness of 6 mm, the other with a rather extreme kerf width of 1.0 mm and a material section of 10 mm.

Table 10.10 reveals that for most cutting applications the losses by convection are small enough to be considered negligible. Once more the only time the losses became appreciable would be during the cutting of thick section thermoset polymers or wood. Even in this case the losses could be expected to be less than the maximum "$h = 1.0$" figures quoted in Table 10.10 and are likely to be of the order of a few tens of watts.

10.6 Notes on the Focusing Characteristics of CO_2 Lasers

10.6.1 Theoretical Focused Spot Size and Depth of Focus

Finding the focal spot diameter of a high power CO_2 laser beam is extremely difficult unless an empirical test such as drilling a series of holes in thin section acrylic is considered accurate enough. The difficulty obviously lies in the very high energy densities available at and near the focal plane. In the absence of good experimental results most manufacturers and researchers refer to the standard optical formulae for focusing a Gaussian distribution beam.

These formulae give

$$\text{Focal point diameter} = \frac{4\lambda f}{\pi D} \quad \text{(Ref. 3, Chap. 11)} \qquad (10.8)$$

$$\text{Depth of focus} = \frac{2\lambda}{\pi}\left[\frac{2f}{D}\right]^2 \quad \text{(Ref. 4, Chap. 11)} \qquad (10.9)$$

Table 10.11. Theoretical focal point diameters and depths of focus

Original beam diameter		Lens focal length		Focal point diameter		Depth of focus	
mm	in	mm	in	mm	in	mm	in
5	0.2	63.5	2.5	0.17	0.0067	4.35	0.171
10	0.4	63.5	2.5	0.086	0.0034	1.09	0.043
15	0.6	63.5	2.5	0.057	0.0022	0.48	0.019
20	0.8	63.5	2.5	0.043	0.0017	0.27	0.011
25	1.0	63.5	2.5	0.032	0.0013	0.17	0.007
5	0.2	127	5.0	0.34	0.0130	17.42	0.686
10	0.4	127	5.0	0.17	0.0067	4.36	0.172
15	0.6	127	5.0	0.11	0.0043	1.94	0.076
20	0.8	127	5.0	0.086	0.0034	1.09	0.043
25	1.0	127	5.0	0.068	0.0027	0.70	0.027
5	0.2	190.5	7.5	0.52	0.0205	38.99	1.535
10	0.4	190.5	7.5	0.26	0.0103	9.74	0.383
15	0.6	190.5	7.5	0.17	0.0067	4.33	0.170
20	0.8	190.5	7.5	0.13	0.0051	2.44	0.096
25	1.0	190.5	7.5	0.10	0.0039	1.56	0.061

where

λ = Wavelength of the light
f = Lens focal length
D = Original unfocused beam diameter

The results for the beam diameters and lens focal lengths in common use given by these formulae are shown in Table 10.11.

10.6.2 The Discrepancy Between Theory and Practice

These theoretical results do not match the actual values of focal spot diameter and depth of focus which can be estimated from the cutting process. Taking a typical example of a 15 mm diameter beam focused by a 127 mm focal length lens, Table 10.11 gives a focal spot diameter of 0.11 mm and a depth of focus of 1.94 mm. Measurement of the minimum kerf width possible when cutting would, however, reveal an effective focal spot diameter closer to 0.25 mm and a depth of focus of approximately 3.0 mm. The idea of an effective focal spot diameter is based upon empirical evidence rather than calculation and gives a more useful measure of the focusing characteristics of a beam. A good test material is thin section (~1 mm) acrylic sheet which has a low thermal conductivity and is readily evaporated by the laser.

The most accurate test is to evaporate a hole from the material whilst the material and the laser are both stationary. The duration of the laser–material interaction should be approximately an order of magnitude larger than that needed to just pierce the material. This ensures that the whole beam is passing through the hole and therefore the hole is just larger than the beam cross section. For example; a 1 s shot from any laser with a power of more than 100 W should be sufficient. The laser beam does not radiate substantial amounts of energy perpendicular to its direction of propagation, and this, together with the low conductivity of acrylic, suppresses any tendency for the drilled hole to expand once the beam has fully penetrated it.

This method gives a more accurate measure of the focusing characteristics of the beam than the theoretical approach. The results of empirical tests of this sort always reveal that the focused spot diameter is always larger than theoretically expected (usually by a factor of 2 or 3). The depth of focus is also larger than theoretically predicted, but this is to be expected if the beam is not being focused down to its smallest possible size. The reasons why the focused spot size is larger than the theory would suggest can be divided into two groups: laser beam considerations and lens imperfections. These will now be discussed separately.

10.6.2.1 Laser Beam Considerations

Equations (10.8) and (10.9) assume that the incident beam has a Gaussian energy distribution. This is never true of a CO_2 laser used for cutting.

Most lasers used for cutting have an energy cross section which gets higher towards the centre, but this can be only truly Gaussian under the best optical conditions and at the lasing threshold (i.e. the internal energy level at which lasing will only just take place). All industrial lasers operate at much higher energy densities, and this leads to a corruption of the mode. Thermal gradients in the electric discharge, vibration, poor quality optics and sometimes deliberate optical design all act together to make the mode non-Gaussian. Non-Gaussian modes do not obey the rules of Eqs (10.8) and (10.9) and do not focus to such a small diameter spot. For example, a uniform energy density or "top hat" mode will focus down to a spot with approximately three times the diameter of the equivalent Gaussian beam. This increase in diameter has strong effect on the energy density of the beam which is proportional to r^2 (a spot of about three times the diameter has an energy density of only ~10% of the smaller focused spot).

Another impractical assumption of the theoretical approach is that the unfocused beam exits as a parallel sided column of light. In practice all high power CO_2 lasers have a divergent beam, although many are designed to converge the beam for the first metre or so of propagation. This divergence or convergence/divergence will have a deleterious effect on the focusing of the beam, particularly on moving optics systems where the changing beam size will affect the focused spot size. Multi-lens collimators can be employed where such an effect gives problems. These are positioned shortly after the beam exits from the laser cavity and ensure a uniform beam cross section.

The final disagreement between theory and practice comes with the assumption that the laser beam is perfectly monochromatic. It was mentioned in Sect. 9.2.2 that a number of energy transitions are available as part of the lasing reaction. Although selective competition between these transitions ensures that the great majority of the photons emitted have a wavelength of 10.6 μm, there will always be a small proportion of the beam which has a slightly higher or lower wavelength. These different wavelength components will be focused to a different focal point than the main body of the beam and will therefore inhibit the generation of the smallest focal spot.

10.6.2.2 Lens Imperfections

Any damage to a lens will make it less capable of focusing the beam down to a minimum diameter spot. Even new lenses have an intrinsic imperfection which limits their performance; the focusing lenses used in laser cutting machines usually consist of a single zinc selenide (or alternative material, e.g. GaAs, KCL) blank which has had one or both faces ground to part of a spherical curve in order to diffract the incident beam towards the focus. Simple lenses of this type are cheap to manufacture, but suffer from spherical aberration which tends to increase the size of the focused spot and the depth of focus. Two methods of combating this phenomenon are possible, but both are expensive in comparison with the use of the single spherical curvature lens. The first method is that which is used in conventional optical apparatus; multiple lens systems. The second is the use of an aspherical lens which has a complex cross sectional profile designed especially to counteract spherical aberration.

10.6.3 Concluding Comments

Although Eqs (10.8) and (10.9) do not give the correct results for the sizes of the focused spot and the depth of focus they do indicate the relationships between the various parameters involved, i.e.:

1. The focal point diameter and the depth of focus both increase as λ or f are increased.
2. The focal point diameter and the depth of focus both decrease if D is increased.

The practical outcome of this is that to get the smallest spot size we need a large diameter incident beam and a short focal length lens. The expansion of the beam can easily be achieved by use of a multi-optic collimator/expander. The expanded beam could then the passed through a short focal length lens which may be aspherical to reduce the spherical aberration and therefore generate the smallest spot.

Apart from the capital cost of such a system there are other factors which explain why multiple optics are not in common use on laser cutting machines:

1. The number of optics in the system should be kept to a minimum as each optic is a potential source of beam corruption if it becomes dirty or scratched. (This is particularly true of transparent optics such as lenses.)
2. Maintenance costs can be high in terms of new optics and production time lost.
3. Each optic will absorb part of the beam and thus reduce the power available at the cut zone.
4. Large diameter short focal length lenses needed to collect and focus the expanded beam suffer from three drawbacks:

   (a) They involve a lot of raw material and work, and are therefore expensive.
   (b) They cannot withstand high cut gas pressures.
   (c) They are necessarily positioned close to the cut zone and are therefore prone to spatter damage etc.
5. The cutting process may not necessarily be improved or accelerated by using a smaller focused spot. Fluid dynamic considerations in the cut zone mean that there will be an optimum kerf width which may not necessarily be the minimum achievable.
6. The reduction in focal spot diameter will be accompanied by a reduction in depth of focus (see Eqs (10.8) and (10.9)). This will make the cutting process less reliable as it will become very sensitive to changes in the lens–workpiece distance as a result of non-flat sheets etc.

As Eqs (10.8) and (10.9) show the relationships between the beam and lens parameters and the focusing characteristics, they could be used to good effect if a correction factor was added. The two equations would then read as follows:

$$\text{Focal point diameter} = \frac{4\lambda f A}{\pi D} \tag{10.10}$$

$$\text{Depth of focus} = \frac{2\lambda B}{\pi}\left[\frac{2f}{D}\right]^2 \tag{10.11}$$

where A and B are experimentally determined values for a particular machine. (The experiment would consist of drilling holes in this acrylic sheet at various positions at and around the focal plane; see Sect. 10.6.2.) The magnitude of A and B for most CO_2 lasers used for cutting would generally lie between 2 and 5.

Chapter 11

Bibliography and Further Reading

The following bibliography gives a short list of selections of published work which could be of interest as further reading. This list is very brief and a great deal of excellent work has had to be omitted in order to keep it short. I apologise in advance to those authors and editors whose work I have left out. A large proportion of work produced on the subject of laser processing is published as part of the proceedings of the conferences which have multiplied over the past few years. Most of these conferences publish their proceedings as a single volume and these are generally useful compendiums of the publications of a number of workers.

After brief notes on some of the major conference proceedings and a selection of journals books and relevant to the subject, there follows a list of specific papers which could form the basis of further reading. The books and papers have been numbered consecutively and these numbers act as the reference numbers used in earlier sections of this book.

11.1 Conference Proceedings and Journals

11.1.1 Conference Proceedings

A. The Lasers in Manufacturing (LIM) Series

LIM1. 1–3 Nov 1983, Brighton, UK (Ed. M.F. Kimmitt)

LIM2. 26–28 March 1985, Birmingham, UK (Ed. M.F. Kimmitt)

LIM3. 3–5 June 1986, Paris, France (Ed. A. Quenzer)

LIM4. 12–14 May 1987, Birmingham, UK (Ed. W.M. Steen)

LIM5. 13–14 September 1988, Stuttgart, Germany (Ed. H. Hugel)

LIM6. 10–11 May 1989, Birmingham, UK (Ed. W.M. Steen)

Conferences organised and proceedings published by IFS Publications and Springer-Verlag. Contact:

IFS (Conferences) Ltd.
35–39 High Street,
Kempston,
Bedford MK42 7BT, UK
Tel. (0234) 853605

The LIM Conferences are a highly successful annual series. Each proceedings volume contains between 20 and 30 papers by a worldwide selection of authors, although there is a natural tendency to concentrate on European work. As the name of the series implies, a large proportion of the work presented involves reports from industrial rather than scientific workers.

B. SPIE Conference Proceedings

International Society for Optical Engineering (SPIE)
P.O. Box 10,
Bellingham, WA 98227, USA
Tel. (206) 676-3290

SPIE is a non-profit technical society dedicated to advancing engineering and scientific applications of optical and laser technology. The society has been active in organising conferences and published proceedings in the laser field since 1969. The content of the proceedings is, naturally, determined by the title of the conference. In general the papers submitted are of a technical or research nature. A few examples of SPIE proceedings which may be of interest are:

Laser Applications in Materials Processing. Proc. Vol. 198 (1979) San Diego CA (Ed. J.F. Ready)

High Power Lasers and Applications. Proc. Vol. 270 (1981) Los Angeles CA (Ed. C.C. Tang)

Applications of High Power Lasers: Critical Review of Technology. Proc. Vol. 527 (1985) Los Angeles CA (Ed. R.R. Jacobs)

High Power Lasers. Proc. Vol. 801 (1987) The Hague, The Netherlands (Eds E.W. Kreutz, A. Quenzer, D. Schuocker)

High Power Co_2 Laser Systems and Applications. Proc. Vol. 1020 (1988) Hamburg, Germany (Ed. A. Quenzer)

CO_2 Lasers and Applications. Proc. Vol. 1042 (1989) (Eds J. Evans, E. Locke)

High Power Lasers and Laser Machining Technology. Proc. Vol. 1132 (1990) (Eds M. Gaillard, A. Quenzer)

The escalation in the volume numbers over the years gives an indication of the amount of material generated by these SPIE conferences. Direct contact with SPIE should unearth a paper relevant to almost any aspect of lasers and laser processing.

C. The ECLAT Conferences

European conferences on Laser Treatment of metals. Organised and published by the German Welding Society:

Deutscher Verband fur Schweisstechnic e.v.
Dusseldorf,
Germany

The contributions to these conferences are generally of a technical–research nature covering high power laser processing techniques.

Early conferences ('86 and '88) were mostly in English. More recently the work has been predominantly in German.

D. The ICALEO Conferences

The annual International Conference on Applications of Lasers and Electro Optics (ICALEO) is organised by the Laser Institute of America. These conferences cover all aspects of laser applications and the published proceedings are a mine of information written by some of the world's top specialists. See also the compilation volumes and other Laser Institute of America publications mentioned in section 11.5.

Details can be obtained from:

Laser Institute of America
12424 Research Parkway,
Suite 130,
Orlando, FL 32826, USA
(Tel. (407) 380-1553)

11.1.2 Journals

E. The Journal of Laser Applications

Publisher: The Laser Institute of America, ISSN 1042-346X

Contact: Journal of Laser Applications,
12424 Research Parkway,
Suite 125,
Orlando, Florida 32826, USA

This journal publishes papers on a wide range of laser related topics. The content is generally divided into four areas:

1. Materials Processing and Interaction
2. Laser sensing and measurement
3. Biomedical applications
4. Laser safety

F. *Lasers in Engineering*

Publisher: Gordon & Breach, ISSN 0898-1507

Contact: Editor, B.L. Mordike
Inst. für Werkstoffkunde und Werkstofftechnik,
Technische Universität Clausthal,
Agricolstrasse 6,
D. 3392,
Clausthal-Zellerfeld,
Germany

This journal publishes papers on all aspects of the application of lasers in industry from cutting, welding and heat treatment to electronic component manufacture and measurement techniques.

11.2 Books

1. The Industrial Laser Annual Handbook
Editors: David Belforte and Morris Levitt
Pennwell Books,
1421 South Sheridan Road,
P.O. Box 1260,
Tulsa, OK 74101, USA

The handbook, which as the name implies is published annually, is divided into several sections:

Laser Materials Processing Data and Guidelines (data tables and curves)

Annual Review of Laser Processing (specific papers by invited authors)

Company and Product Directory (international company review and product specifications) Related Products and Services (spares, optics etc.)

The subject matter includes review of the world market and welding, cutting, heat treatment, etc., using CO_2, Nd:YAG and eximer lasers. The publication is related to the popular professional magazine *Laser Focus* with which it shares its excellent data base on companies and services.

2. W.W. Duley. Laser Processing and Analysis of Materials. Plenum Press, New York, 1983 (ISBN 0-306-41067-2)

This general textbook contains over 400 pages of information on a wide variety of lasers and their applications. The lasers described range from milliwatt diode lasers to multikilowatt CO_2 devices. Methods of processing cover a similarly large scope from the remote sensing of water quality to isotope separation. The breadth of the work makes it a worthwhile addition to any laser researcher's library, but the majority of techniques described are related more to the laboratory than the industrial shop floor.

3. W.M. Steen. Laser Material Processing. Springer-Verlag, Berlin Heidelberg New York, 1991 (ISBN 3-540-19670-6 (UK); ISBN 0-387-19670-6 (USA))

This book covers the whole field of laser material processing and includes chapters on laser optics, cutting, welding, heat flow theory, surface treatment, automation and safety. This provides useful background to the subject as a whole and is another important source book. The style is an easily readable combination of conversational and technical.

4. J. Bellis (ed.). Lasers; Operation, Equipment, Application and Design. Prepared by the engineering staff of Coherent Inc. McGraw Hill, 1980 (ISBN 0-07-011593-1)

A non-scientific coverage of a large range of commercial applications aimed at the potential customer. A clear, well presented description of various aspects of laser physics and design is followed by almost one hundred pages of industrial examples covering applications from processing of plastics to heat treatment of metals.

5. E.A. Metzbower (ed.). Source Book on Applications of the Laser in Metal Working. American Society for Metals, 1981

An excellent and varied collection of papers on all aspects of laser applications in the metal working industries including nine articles on cutting, three on drilling, eight on surface hardening etc.

6. M. Bass (ed.). Laser Materials Processing. North Holland, Amsterdam, 1983

A collection of specially prepared chapters by different authors which includes almost one hundred pages on the subject of laser cutting by W.M. Steen and J. Kamalu.

7. J.F. Ready (ed.). Industrial Applications of Lasers. Academic Press, New York, 1978 (ISBN 0-12-583960-X)

Although this book is now rather out of date for a work on this subject, it does contain a great deal of useful information and interesting discussion.

8. A.L. Bloom (ed.). Gas Lasers. John Wiley (University of California Berkeley, Letters and science extension series)

A clear and detailed description of the physics of gas lasers for the engineer or scientist.

11.3 Specific Papers

The following sections list a few of the best specific papers in English, covering various aspects of CO_2 laser materials processing, particularly cutting. This list is not intended to be an exhaustive survey but acts as a starting point of a literature search. For reasons of economy of space, many good pieces of work have been omitted from the compilation and I would like to apologise to the various authors concerned. The papers mentioned below have been arranged in alphabetical order of the first author's name in the following subsections:

11.3.1 General reviews of laser processing and cutting
11.3.2 Cutting metals
11.3.3 Cutting non-metals
11.3.4 Theoretical analyses of the cutting process
11.3.5 CO_2 lasers
11.3.6 Nozzle design and gas dynamics
11.3.7 Laser beam analysis
11.3.8 Alternative cutting methods

11.3.1 General Reviews of Laser Processing and Cutting

9. D.A. Belforte. Laser applications in the auto industry. In: H. Hugel (ed.). Proc. 5th Int. Conf. Lasers in Manufacturing (LIM 5) Stuttgart, Germany, 13–14 September 1988, pp 61–70
10. L. Hanicke. Laser technology in the Volvo Car Corporation. In: W.M. Steen (ed.). Proc. 4th Int. Conf. Lasers In Manufacturing (LIM 4) Birmingham, UK, 12–14 May 1987, pp 49–68
11. J. Powell, I.A. Menzies, P.F. Scheyvaerts. Laser cutting: the state of the art. In: Proc. 2nd Eur. Conf. on Laser Treatment of Materials (ECLAT '88), Bad Nauheim, Germany, 13–14 October 1988, pp 1–9
12. D.M. Roessler. U.S. automotive applications of laser processing. In: D. Belforte, M. Levitt (eds). The Industrial Laser Annual Handbook, 1988, pp 75–83

13. A.J. Schwarz, J. Powell. Laser cutting jobshops; a technical and commercial appraisal. In: W.M. Steen (ed.). Proc. 6th Int. Conf. Lasers in Manufacturing (LIM 6), Birmingham, UK, 10–11 May 1989, pp 173–181

11.3.2 Cutting Metals

14. Y. Arata, H. Maruo, I. Miyamoto, S. Takeuchi. Improvement of cut quality in laser-gas-cutting stainless steel. In: Proc. 1st Int. Laser Processing Conf., Anaheim, CA, 16–17 November 1981. Laser Institute of America
15. Y. Arata, H. Maruo, I. Miyamoto, S. Takeuchi. Dynamic behaviour in laser gas cutting of mild steel. Transactions of the Japanese Welding Research Institute 1979; 8(2): 15–26
16. G. Daurelio, M. Dell'Erba, L. Cento. Cutting copper sheets by CO_2 laser. Lasers and Applications 1986; March: 59–64
17. J.T. Gabzdyl, W.M. Steen, M. Cantello. Nozzle beam alignment for laser cutting. In: Proc. ICALEO '87 San Diego CA, USA, May 1987, LIA Toledo, 1988, pp 143–148
18. M. Hansmann, I. Decker, J. Ruge. Influence of workpiece temperature in laser cutting. In: Proc. 1st European Conference on Laser Treatment of Materials. (ECLAT 86) Bad Nauheim, Germany, 1986
19. You-Hie Han, I. Decker, J. Ruge. Laser cutting of selected steels: micro structural changes and their effect on mechanical properties. In: Proc. 1st Eur. Conf. on Laser Treatment of Materials, Bad Nauheim, 1986. German Welding Society (DVS)
20. A. Ivarson, J. Powell, C. Magnusson. The role of oxidation in laser cutting stainless and mild steel. Journal of Laser Applications 1991; 3(3): 41–45
21. A. Ivarson, J. Powell, C. Magnusson. Laser cutting of steels: analysis of the particles ejected during cutting. Welding in the World 1992; 30(5/6): 116–125
22. A. Ivarson, J. Powell, G. Broden, J. Kamalu, C. Magnusson. The effects of oxygen purity in laser cutting mild steel: a theoretical and experimental investigation. Lasers in Engineering 1992/93
23. V.S. Kovalenko, Y. Arata, H. Maruo, I. Miyamoto. Experimental study of cutting different materials with a 1.5 kW CO_2 laser. Transactions of the Japanese Welding Research Institute (JWRI) 1978; 79(2): 101–112
24. T. Nakagawa, K. Suzuki, K. Sakaue. Laser cut blanking tool. Bulletin of the Japan Society of Precision Engineering 1983; 17(1): 45–46 (Also available as: Manufacturing of a blanking tool by laser machining. Report published by Coherent (8/82.5M 5055A))
25. F.O. Olsen. Studies of sheet metal cutting with plane polarised CO_2 laser. In: Proc. 5th Annual Congress on Opto Electronics in Engineering (Laser 81), Munich, W. Germany, pp 227–231
26. D. Petring, D. Abels, E. Beyer. Absorption distribution on idealised cutting front geometries and its significance for laser beam cutting. In: High Power CO_2 Laser Systems and Applications 19–20 September 1988, Hamburg, Germany. Proc. SPIE Vol. 1020, pp 123–131
27. D. Petring, E. Beyer. Laser beam cutting of stainless steel. Welding in the World 1993 (to appear) (IIW Document No. 1E 116–91)
28. D. Petring, P. Abels, E. Beyer, W. Noldechen, K.U. Preissig. Laser beam cutting of highly alloyed thick section steels. In: Proc. Laser 1989, pp 599–604
29. J. Powell, K. Frass, I.A. Menzies, H. Fuhr. CO_2 laser cutting of non ferrous metals. In: High Power CO_2 Laser Systems and Applications, Hamburg 19–20 September 1988. Proc. SPIE Vol. 1020, pp 156–163
30. J. Powell, K. Frass, I.A. Menzies. The influence of laser mode quality on cutting performance. In: Proc. Laser 4, Versailles, France, 17–18 March 1988
31. J. Powell, A. Ivarson, L. Ohlsson, C. Magnusson. Conductive losses experienced during CO_2 laser cutting. Journal of Laser Applications 1993 (to appear)
32. J. Powell, T.G. King, I.A. Menzies. Cut edge quality improvement by laser pulsing. In: 2nd Int. Conf. on Lasers in Manufacturing (LIM 2), Birmingham, UK, 26–28 March 1985, pp 37–45
33. J. Powell, T.G. King, I.A. Menzies, K. Frass. Optimisation of pulsed laser cutting of mild steels. In: 3rd Int. Conf. on Lasers in Manufacturing (LIM 3) Paris, France, 3–5 June 1986, pp 67–75
34. J. Powell, I.A. Menzies. Metallurgical implications of laser cutting stainless steels. In: Proc. 1st Int. Conf. on Power Beam Technology, Brighton, UK 10–12 September 1986. Welding Institute

35. J. Powell, A. Ivarson, L. Ohlsson, C. Magnusson. Reflection phenomena in CO_2 laser cutting. Journal of Laser Applications 1993 (to appear)
36. J. Powell, P.F. Scheyvaerts, K. Frass, I.A. Menzies. The influence of focal point penetration on cutting dynamics for thick section steels at a CO_2 laser power of 2.5 kW. In: Proc. 2nd Eur. Conf. on Laser Treatment of Materials (ECLAT '88), Bad Nauheim, Germany, 13–14 October 1988, pp 114–118
37. J. Powell, M. Jezioro, I.A. Menzies, P.F. Scheyvearts. CO_2 laser cutting of titanium alloys. In: Proc. Int. Conf. Laser Technologies in Industry, Porto, Portugal, 6–8 June 1988
38. W. Schulz, G. Simon, H. Urbassek, M. Decker. On laser fusion cutting of metals. Journal of Physics D: Applied Physics 1987; 20: 481–488

11.3.3 Cutting Non-metals

39. D.J. Doyle, J.M. Kokosa. Hazardous by products of plastics processing with CO_2 lasers. In: Int. Congress on Applications of Lasers and Electro Optics (ICALEO 1985) San Francisco, CA, 11–14 November 1985
40. M. Flaum, T. Karlson. Cutting of fibre reinforced polymers with a c.w. CO_2 laser. In: E.W. Kreutz, A. Quenzer, D.S. Schuocker (eds). Proc. High Power Lasers, The Hague, The Netherlands, 31 March–3 April 1987. Proc. SPIE, Vol. 801, pp 130–137
41. R. Nuss, R. Muller, M. Geiger. Laser cutting of RRIM–Polyurethane components in comparison with other cutting techniques. In: Proc. 5th Int. Conf. on Lasers in Manufacturing (LIM 5), Stuttgart, 1988, pp 47–57
42. S.E. Nielsen. Laser material processing of polymers. Polymer Testing (GB) 3(4): 303–310
43. J. Powell, G. Ellis, I.A. Menzies, P.F. Scheyvaerts. CO_2 laser cutting of non metallic materials. In: W.M. Steen (ed.). Proc. 4th Int. Conf. Lasers in Manufacturing (LIM 4), Birmingham UK, 12–14 May 1987, pp 69–82
44. J. Powell, G. Ellis, C.D. Young, I.A. Menzies. Laser cutting of polymeric materials: an experimental investigation. In: Proc. Int. Conf. on Laser Materials Processing – Science and Applications, Osaka, Japan, 21–23 May 1987
45. J. Powell, I.A. Menzies, G. Ellis, P.F. Scheyvaerts. High quality cutting of non metals by CO_2 laser. In: D. Belforte, M. Lewitt (eds). The Industrial Laser Handbook, 1988 edition, pp 56–60

11.3.4 Theoretical Analyses of the Cutting Process

46. Y. Arata, I. Miyamoto. Some fundamental properties of high power laser beam as a heat source (report 2) – CO_2 laser absorption characteristics of metal. Transaction Japanese Welding Society 1972; 3(1): 152–162
47. Y. Arata, M. Kanayama. Analysis of thermal cutting based on heat conduction theory. In: Proc. 2nd Colloquium on Electron Beam Welding and Melting (2nd CISFFE), Avignon, 5–8 September 1978, pp 429–434
48. K.C.A. Crane, R.K. Garnsworthy, L.E.S. Mathias. Ablation of materials subjected to laser radiation and high speed gas flows. Journal of Applied Physics 1980; 51(11): pp 5954–5961
49. I. Decker, J. Ruge, U. Atzert. Physical models and technological aspects of laser gas cutting. In: Conf. on Industrial Applications of High Powered Lasers, 26–27 September 1983 Linz, Austria. Proc. SPIE Vol. 455, pp 81–87
50. F.O. Olsen. Theoretical investigation in the fundamental mechanisms of high intensity laser light reflectivity. In: High Power CO_2 Laser Systems and Applications, 19–20 September 1988 Hamburg, W. Germany. Proc. SPIE Vol. 1020, pp 114–122
51. J. Powell. The influence of material thickness on the efficiency of laser cutting and welding. In: Proc. 6th Int. Conf. on Lasers in Manufacturing (LIM 6), Birmingham, UK, 10–11 May 1989, pp 215–221
52. W. Schulz, G. Simon, M. Vicanek. Ablation of opaque surfaces due to laser irradiation. Journal of Physics D: Applied Physics 1986; 19: 173–177
53. M. Vicanek, G. Simon, H.M. Urbassek, I. Decker. Hydrodynamical instability of melt flow in laser cutting. Journal of Physics D: Applied Physics 1987; 20: 140–145

11.3.5 CO_2 Lasers

54. O. Maerten, G. Hertzinger, P. Klein, P. Loosen. Laser–medium/resonator field interaction of fast axial flow CO_2 lasers. In: E.W. Kreutz, A. Quenzer, D.S. Schuocker (eds). High Power Lasers, The Hague, The Netherlands, 31 March–3 April 1987. Proc. SPIE Vol. 801, pp 51–57
55. C.K.N. Patel. High power carbon dioxide lasers. Scientific American 1968; August: 23–33
56. A.L. Schawlow. Advances in optical masers. Scientific American 1963; July
57. A. Sona. R.F. and D.C. excited high power lasers. In: E.W. Kreutz, A. Quenzer, D.S. Schuocker (eds). High Power Lasers, 31 March–3 April 1987. Proc. SPIE Vol. 801, pp 23–31

11.3.6 Nozzle Design and Gas Dynamics

58. B.A. Ward. Supersonic characteristics of nozzles used with lasers for cutting. In: Proc. Int. Conf. on the Application of Lasers and Electro Optics. (ICALEO 84) Boston, MA, 12–15 November 1984
59. J. Fieret, M.J. Terry, B.A. Ward. Aerodynamic interactions during laser cutting. In: Int. Symposium on Optical and Opto electronic Applied Sciences and Engineering, Conference on Laser Processing: Fundamentals, Applications and Systems Engineering, Quebec, Canada, 2–6 June 1986. Proc. SPIE
60. J. Fieret, B.A. Ward. Circular and non-circular nozzle exits for supersonic gas jet assist in CO_2 laser cutting. In. A. Quenezer (ed.). Proc. 3rd Int. Conf. Lasers in Manufacturing (LIM 3), June 1986, Paris, France, 3–5 June 1986, pp 45–54
61. J. Fieret, M.J. Terry, B.A. Ward. Overview of flow dynamics in gas assisted laser cutting. In: E.W. Kreutz, A. Quenzer, D.S. Schuocker (eds). Proc. High Power Lasers The Hague, The Netherlands, 31 March–3 April 1987. SPIE Vol. 801, pp 243–250
62. S.E. Neilson. Laser cutting with high pressure cutting gases and mixed gases. In: A. Quenzer (ed.). Proc. 3rd Int. Conf. Lasers in Manufacturing (LIM 3) Paris, France, 3–5 June 1986, pp 25–44
63. A.E. Puckett. Supersonic nozzle design. Journal of Applied Mechanics 1946; December: A265–A270

11.3.7 Laser Beam Analysis

64. G.C. Lim, W.M. Steen. The measurement of the temporal and spatial power distribution of a high powered CO_2 laser beam. Optics Laser Technology 1982; June: 149–153
65. G.C. Lim, W.M. Steen. Instrument for the instantaneous in-situ measurement of a high power CO_2 laser beam. Journal of Physics E: Scientific Intrumentation 1984; 17: 999–1007
66. I. Miyamoto, H. Maruo, Y. Arata. Intensity profile measurement of focused CO_2 laser beam using PMMA. Proc. ICALEO 1984. Laser Institute of America, Vol. 44, pp 313–320
67. G. Sepold, P.O. Juptner, J. Telepski. Measuring the quality of high power laser beams. In: High Power Lasers and their Industrial Applications, Innsbruck, Austria, 15–18 April 1986. Proc. SPIE Vol. 650, pp 167–169
68. B.A. Ward. Laser beam measurement and quality assurance. In: 2nd Symposium on Exploiting the Laser in Engineering Production, Coventry, UK, September 1984. The Welding Institute, Cambridge

11.3.8 Alternative Cutting Methods

69. W. Dobbins, H. van Arb. Using the "slab" laser. The Fabricator 1990; December
70. M. Hashish. Application of abrasive water jets to metal cutting. In: Proc. Conf. Non Traditional Machining, Ohio, IL, USA, December 1985
71. M. Hashish. Aspects of abrasive water jet performance optimisation. In: Proc. 8th Int. Sym. on Jet Cutting Tech. Durham, UK, 9–11 September 1986, pp 297–308
72. T.J. Labus, R. Pilarski. Fluid jet technology for industrial applications. In: Proc. Conf. Non Traditional Machining, Ohio, IL, USA, December 1985

73. M. Mawson. Mechanised plasma arc cutting. In: Proc. 1st Int. Conf. on Developments and Innovations for Improved Welding Production, Birmingham UK, September 1983. Welding Institute, Cambridge
74. L. Ohlsson, J. Powell, A. Ivarsson, C. Magnusson. Comparison between abrasive water jet cutting and laser cutting. Journal of Laser Applications 3(3): 46–50
75. J. Powell, C. Wykes. A. comparison between CO_2 laser cutting and competitive techniques (Nd:YAG Laser, Plasma Arc, Abrasive Water Jet and Oxygen Flame). In: W.M. Steen (ed.). Proc. 6th Int. Conf. on Lasers in Manufacturing, Birmingham, UK, May 1989, pp. 135–153
76. Industrial Processing Applications: Raytheon Nd:YAG Lasers. Raytheon Laser Centre, Burlington MA. 01803 USA
77. Welding cutting and related processes. In: Welding Handbook, Section 3A. A.L. Phillips (ed.). American Welding Society
78. Nd:YAG Laser Applications Data. Lumonics, Rugby, UK

11.4 Miscellaneous References

79. R.W. Weast (ed.). Handbook of Chemistry and Physics, 1st Student Edition, CRC Press, Boca Raton, FL, 1987 (ISBN 0-8493-0740-6)
80. R.M. Tennent (ed.). Science Data Book, Oliver and Boyd, Edinburgh, 1986 (ISBN 0-05-002487-6)
81. W. Bolton. Newnes Engineering Materials Pocket Book, Heinemann-Newnes, Oxford, 1990 (ISBN 0-434-90113)
82. I. Barin, Knacke. Thermochemical Properties of Inorganic Substances. (ISBN 3-514-00125-1)
83. G.F.C. Rogers, Y.R. Mayhew. Engineering Thermodynamics, Work and Heat Transfer, 3rd edn. Longman Scientific and Technical, Harlow, 1980 (ISBN 0-582-30500-4)

11.5 Further reading update for the second edition of this book

Books and Paper Collections:

84. Laser cutting and drilling reference guide. (Paper collection)
pub: Laser Institute of America
12424 Research Parkway,
Suite 125,
Orlando, Florida 32826, USA
A valuable collection of 22 papers on laser cutting and 9 papers on drilling taken from the Journal of Laser applications and the ICALEO conferences.
85. Laser Cutting Guide (book)
Author: John Powell
Publisher: Laser Institute of America
12424 Research Parkway,
Suite 125,
Orlando, Florida 32826, USA
To be published late in 1998.
A booklet covering the basics of CO_2, Nd:YAG and Eximer laser cutting aimed at readers with only moderate technical education.

86. L.I.A. Laser Materials Processing Handbook
Editors: Jack F. Ready and Dave F. Farson
Publisher: Laser Institute of America and Magnolia Publishing Inc.
12424 Research Parkway,
Suite 125,
Orlando, Florida 32826, USA
ISBN 0-912035-15-3
To be published late 1998
A 450 page handbook covering all aspects of laser materials processing; laser systems, safety, heat treatment, welding, cutting, drilling, marking, rapid prototyping, electronic materials processing etc.
87. Laser Materials Processing (book)
Editor: L. Migliore
pub: Marcel Dekker Inc. New York USA 1996
ISBN 0-8247-9714-0
88. On the Physics and Chemical Thermodynamics of Laser Cutting (Ph.D. thesis)
Author: Anders Ivarson
pub: Lulea University, Sweden 1993
ISSN 0348-8373
A collection of six papers (written in English) covering some of the material from chapter 10 of this book in greater detail. Additional work includes "Factors affecting the efficiency of inert gas laser cutting" and "Optimisation of the laser cutting process for thin section stainless steels".

Individual papers

89. W. O'Neill, W.M. Steen. A three dimensional analysis of gas entrainment operating during the laser cutting process. Journal of Physics; D. Applied Physics v.28. Jan 14, 1995 pp 12–18
90. G. Caprino, V. Tagliaferri and L. Covelli. The Importance of material structure in the laser cutting of glass fibre reinforced plastic composites. Journal of Engineering Materials and Technology v 117, Jan 95 pp 133–8
91. H.O. Ketting, F.O. Olsen. Laser cutting with different laser sources. Welding in the World vol 37 No 6 Nov–Dec 1996 pp 288–292
92. T. Kristensen, F. Olsen. Investigation of cutting Al alloys with pulsed and c.w. CO_2 lasers. Welding in the world vol 33 No 5 Sept–Oct 94 pp 355–361
93. M.J. Hsu, P.A. Molian. Off axial gas jet assisted laser cutting of 6.35mm thick stainless steel. Journal of Engineering for Industry v 117 May 95 pp 272–6
94. L. Cai, P. Sheng. Analysis of laser evaporative and fusion cutting. Journal of Manufacturing Science and Engineering v118 May 96 pp 225–34
95. A.P. Hoult. Laser cutting and drilling of composites. Industrial Laser Review, August 95 pp 13–16
96. Norikazu Tabata, Shigenori Yagi, Masao Hishii. Present and Future of Lasers for fine cutting of Metal plate. Journal of Materials Processing Technology vol 62 (1996) pp 309–314
97. J. Powell, H. Haferkamp, F.W. Bach, D. Seebaum and A. Hampe. Filtration of the fumes generated by industrial CO_2 laser cutting. Journal of Laser Applications (1996) 8. 241–246
98. W. O'Neill, W.M. Steen. A review of theoretical models of Laser Cutting. Lasers in Engineering vol 3 1994 pp 281–299
99. W. O'Neill, J. Gabzdyl. The mass transfer behaviour of gas jets in laser cutting. Welding in the World v35 No 1 1995 pp 6–11
100. K.U. Preissig, D. Petring, G. Herziger. High Speed cutting of thin metal sheets. Proc. Europto series. Conf. Laser Materials processing. 5–8 April 1994 Vienna, Austria. S.P.I.E. vol 2207 pp 96–110
101. A.F.M. Kaplan. An analytical model of metal cutting with a laser beam. J. Applied Physics vol 79 No 5 pp 2198–2208 ('96)
102. A.F.H. Kaplan, O. Wangler and D. Schuocker. Laser cutting: fundamentals of the periodic striations and their on-line detection. Lasers in Engineering vol 6 pp 103–126 ('97)

Index